The Atomic Portrait

How Cryo-Electron Microscopy Brought Life's Molecules Into Focus

Rosalind K. Thayer

Published by NobleTrex Press

For permissions and other inquiries, write to:
P.O. Box 3132, Framingham, MA 01701, USA

Contents

Introduction

Suppose I told you that some of the most important events in your life depend on shapes too small for light itself to reveal.

Whether a virus can invade a cell. Whether a neuron can pass a signal. Whether a cancer drug will fit its target or miss by the width of a misplaced atom. Whether your body's protein-making factories stall, accelerate, or make a fatal mistake. Again and again, life comes down to tiny structures folding, flexing, docking, and snapping apart in a realm so small that color disappears, ordinary microscopes fail, and the word *seeing* starts to sound almost philosophical.

And yet, over the past few decades, scientists have learned to look there.

Not by shrinking themselves in some science-fiction contraption, and not by prying molecules apart to infer their shape from fragments, but by freezing the machinery of life in ultrathin glassy ice and examining it with beams of electrons. The result is one of the most astonishing achievements in modern science: we can now make portraits of molecules with enough detail to trace the coils of proteins, the loops of RNA, the pockets where medicines bind, sometimes even the positions of individual atoms. We can catch molecular machines in mid-action. We can compare one functional state to another as though paging through a flipbook of chemistry.

This was not supposed to be easy. For a long time, many experts thought it might not be possible at all.

Biology presents a special kind of invisibility problem. Living matter is made mostly of light atoms, especially carbon, nitrogen, oxygen, and hydrogen. These are poor scatterers compared with the heavy metals that make traditional electron microscopy images pop. Biological samples are also fragile. Shine too much energy at them and they burn, buckle, charge up, or simply fall apart. Take away the water and they collapse. Let the water freeze the usual way and it forms crystals that shove molecules around and swamp the image with unwanted structure. Even if the sample survives, the individual pictures are often appallingly faint: ghostly silhouettes floating in noise.

Cryo-electron microscopy—cryo-EM for short—solves this by embracing a wonderfully strange compromise. It uses electrons, because electrons have wavelengths short enough to capture atomic-scale detail. But it uses them sparingly, because the beam damages what it reveals. It keeps samples in a vacuum, because electrons demand it. Yet it also preserves them in water, because biology demands that. The trick is to cool the water so quickly that it does not crystallize. Instead, it becomes a glass-like solid, locking molecules in place so fast that they keep something very close to their native forms. Jacques Dubochet and colleagues made that preservation strategy practical in the 1980s. Joachim Frank developed the statistical logic needed to recover structure from countless blurry molecular views. Richard Henderson showed that electron microscopy could, in principle and in practice, reach near-atomic detail for biological molecules. When the 2017 Nobel Prize in Chemistry recognized these three figures, it honored not merely a technique but a new way of knowing life.

That phrase—*knowing life*—is worth lingering over. We often talk as though biology were a list of components: genes, proteins, membranes, organelles. But biology is not a parts catalog. It is behavior arising from shape and motion. A protein is not important simply because it exists; it matters because it folds into a specific form, because one surface carries an electrical pattern that attracts a partner, because a hinge opens at one moment and shuts at another, because a buried amino acid nudges a neighboring helix and changes the fate of a cell. Structure is not decoration. Structure is mechanism.

This is why cryo-EM matters far beyond a specialist's laboratory. If you have ever taken a medicine designed to fit a molecular target, you have benefited from structural biology. If you followed the race to understand SARS-CoV-2, you saw structural biology becoming headline news. The coronavirus spike protein, that crown-like protrusion studding the virus surface, was not just an icon of the pandemic. Its shape—especially the form it adopts before fusing with our cells—was central to vaccine design. Cryo-EM helped make that shape visible quickly and credibly. It helped confirm how to stabilize the spike in the prefusion form that the immune system most needed to see. The elegance of those images was not ornamental. It had consequences measured in doses manufactured, trials launched, and lives saved.

Even outside moments of crisis, the ability to visualize molecular form changes the kinds of questions scientists can ask. Why does one mutation cause a devastating disease while another, only a few atoms away, has little effect? How does a ribosome read genetic instructions and build a protein chain with astonishing fidelity? How do membrane proteins—those elusive gatekeepers embedded in oily cell membranes—open, close, and signal? How do giant viral shells assemble? How do cellular machines avoid tangling with the thousands of other macromolecules crammed into the same microscopic space? A high-resolution structure does not answer every question, but it transforms vague speculation into testable ideas. It turns *something happens here* into *this loop contacts that pocket, this residue stabilizes that state, this hinge must move for the machine to work.*

There is also, if we are honest, a deeper reason this subject exerts such power. Humans are visual thinkers. We trust what we can picture. For centuries, science has repeatedly enlarged the domain of the visible: Galileo turned the telescope skyward and replaced points of light with worlds; microscopes revealed cells, then bacteria, then the intricate anatomy of tissues. Cryo-EM extends that tradition into a scale where seeing becomes partly experimental, partly mathematical, and entirely revelatory. The final image is not a snapshot in the casual sense. It is built from thousands or millions of particle images, each weak and noisy, aligned and averaged by algorithms sophisticated enough to separate signal from illusion. But that does not make the portrait less real. If anything, it makes it more impressive. Reality at this scale does not simply present itself. It has to be coaxed into focus.

And what a peculiar act of coaxing it is.

A modern cryo-electron microscope is among the most exacting scientific instruments ever built. It loathes vibration, drift, dirt, and thermal fluctuation. It requires a high vacuum and intense electrical stability. It relies on exquisitely controlled electron optics to guide a beam through magnetic lenses. The sample rests at cryogenic temperatures on a support only a whisper thick. The camera—today, usually a direct electron detector rather than the older scintillator-based systems—does something that changed the field completely: it records not just an image but a fast sequence of frames, effectively a movie of the sample during exposure. That allows researchers to correct for beam-induced motion, one of the great image-blurring villains of the field. Add powerful computation, statistical methods that are humble enough to guard against overfitting, and data pipelines capable of handling torrents of information, and cryo-

EM starts to look less like a single invention than like a carefully orchestrated alliance among physics, chemistry, engineering, and computer science.

That alliance arrived late, and then all at once.

For years, electron microscopists could produce tantalizing pictures of large biological objects—viruses, filaments, membranes—but many targets remained stubbornly blurry. The field had heroes, insights, and technical progress, but it also had limits that seemed structural. Then came what researchers now call the resolution revolution. Better detectors. Better motion correction. Better statistics. Faster computation, especially with graphics processors. Improved sample supports. More reliable automation. Each advance amplified the others. Images that once suggested lumpy blobs gave way to maps in which alpha helices became unmistakable, beta sheets crisp, side chains traceable. Structural biology changed accordingly. Cryo-EM moved from a method people admired at a distance to one they planned around from the start.

That shift altered scientific culture as much as scientific capability. Laboratories that had spent decades trying to crystallize difficult proteins suddenly had another option. Huge complexes that resisted other methods became tractable. Flexible molecules, once dismissed as untidy troublemakers, could sometimes be separated into distinct conformational states computationally. Public databases filled with maps and models. The visual language of molecular life grew richer. So did the ambitions of the people studying it.

Still, if cryo-EM has a fairy-tale aura from a distance—freeze, image, reconstruct, admire—the reality is gloriously messier. Molecules prefer some orientations and avoid others. Ice can be too thick, too thin, contaminated, crystalline, or mechanically unstable. Samples can aggregate, denature at the air-water interface, or refuse to behave consistently from grid to grid. The beam can make the specimen twitch. The microscope can perform magnificently for one data set and maddeningly for the next. The reconstruction software can classify particles into meaningful states—or into nonsense, if used carelessly. Validation is not bureaucratic cleanup at the end. It is the immune system of the whole enterprise, constantly asking whether the beautiful structure on the screen corresponds to the molecule in the ice or merely to the hopes of the person processing the data.

That tension between beauty and skepticism gives cryo-EM much of its intellectual character. It is a field that rewards technical bravado but punishes self-deception. It depends on engineering solutions, yet some of its deepest ideas are statistical and conceptual. It can produce images of breathtaking clarity, but only by respecting uncertainty at every step. Even the word *resolution* can mislead if we hear it too casually. Higher resolution is not just a sharper picture. It is access to more trustworthy chemistry, firmer mechanisms, stronger inferences about how nature works.

The pages ahead are about that remarkable combination of fragility and precision. They begin where all good stories about seeing must begin: with scale. There is a world smaller than color, smaller than anything a classical optical microscope can cleanly resolve, and life runs on it. From there, the focus tightens on why electrons make such effective probes and why they are also dangerous ones. We will linger over a deceptively simple material—water—and over the astonishing fact that whether it freezes into crystals or vitrifies into glass can determine whether a biological truth survives long enough to be imaged.

We will step inside the microscope itself, a machine that seems almost temperamental in its hatred of disturbance. We will watch cameras become smart enough and fast enough to rescue information that earlier generations lost. We will see how single-particle analysis turns innumerable noisy two-dimensional projections into three-dimensional maps, and how those maps become atomic models only when interpreted with discipline. We will revisit the breakthroughs that made the modern era possible and meet the people whose insights changed what could be seen. We will also look at the molecular celebrities of the field—ribosomes, membrane proteins, viruses—and at the less glamorous but equally consequential infrastructure of public databases, software ecosystems, and shared standards.

And because no account of cryo-EM today would be complete without it, we will spend time with the coronavirus spike, perhaps the most publicly consequential protein shape of the early twenty-first century. It offers a rare case where the abstract triumphs of structural biology met urgent global need in real time.

But the story does not end with polished portraits of isolated molecules. One of the most exciting frontiers lies in seeing molecules in their crowded native neighborhoods: inside cells, among membranes, fibers, vesicles, and other machines. Cryo-electron tomography pushes in that direction, sacrificing some simplicity to gain context. At the same time, new algorithms are learning to pick particles, denoise images, classify heterogeneity, and guide microscopes with increasing autonomy. The field is moving from heroic manual craftsmanship toward something more scalable, more democratic, and perhaps more surprising.

That prospect carries a pleasing irony. The better we become at seeing life's machinery, the more vividly we appreciate that it is not machinery in the coarse industrial sense. It is restless, probabilistic, soft, and contingent. Molecules breathe. Complexes fluctuate. Structures are not statues but populations of possibilities. Cryo-

EM has flourished not because it forces biology into rigidity, but because it has become good at honoring that subtlety while still extracting form from noise.

A portrait, after all, is never merely a copy of a face. A good portrait tells you what kind of presence you are looking at. Cryo-EM has done something analogous for molecular biology. It has not just supplied prettier pictures. It has changed what counts as an answer. It has let scientists ask not only what a molecule is made of, but how it presents itself to partners, how it shifts under stress, how it invites intervention, and how its geometry constrains its fate. At atomic scale, form is not surface. Form is story.

And for the first time in history, we can read much of that story directly.

Chapter 1

The Invisible Kingdom

The drama of cryo-electron microscopy begins with a simple frustration: the most important actors in biology are too small for our eyes, and even for many microscopes, to see clearly. Here the reader enters a realm where color disappears, wavelength becomes destiny, and scientists must coax meaning from whispers of signal without destroying the fragile molecules they hope to observe.

1.1 Beyond the Rainbow: Where Molecules Slip Out of Sight

A living cell is often drawn as a tidy circle with a few organelles floating inside, a kind of biological snow globe. In reality it is closer to a crowded, working city. There are districts with different industries—power generation in mitochondria, shipping and receiving at the cell membrane, manufacturing on ribosomes, waste management in lysosomes. Traffic flows constantly. Messages are copied, edited, and dispatched. And just as a city's most consequential events can turn on a tiny detail—a key in a lock, a signature on a form, a crack in a bridge joint—the cell's decisive moments often hinge on structures far too small to be seen directly by ordinary light.

The natural unit for this world is the nanometer. A nanometer (nm) is a billionth of a meter: $1\,\text{nm} = 10^{-9}\,\text{m}$. That prefix, *nano*, has been worn down by marketing until it sounds like a synonym for "tiny." In the cell it has a more precise meaning. A nanometer is the scale of a protein's working surfaces, the thickness of a lipid membrane, the diameter of the DNA double helix. When a protein grips another protein, it is matching ridges and pockets that can differ by less than a nanometer. When a virus recognizes a host cell, the fit is negotiated by protrusions and grooves measured in a handful of nanometers. A nanometer is not just a small distance; it is the stage on which biological identity is enforced.

One way to make these sizes feel physical is to take a familiar length and keep shrinking it by factors of ten, like walking down a staircase where each step is an order of magnitude. A millimeter is about the thickness of a credit card. Divide it by ten and you get a tenth of a millimeter: you are now in the realm of a human hair's diameter, roughly 70 to $100\,\mu\text{m}$ (micrometers). Divide by ten again and you reach $10\,\mu\text{m}$, the scale of a typical animal cell. A red blood cell is about 7 to $8\,\mu\text{m}$ across: wide enough that, with a basic light microscope, you can see it as a crisp disc and even notice its central pallor.

Drop another factor of ten, and you are at $1\,\mu\text{m}$. Many bacteria live here. *E. coli*, a workhorse of biology labs, is a rod about $1\,\mu\text{m}$ wide and $2\,\mu\text{m}$ long. A good light microscope can show its outline, sometimes even the hint of a dividing septum, but not the machinery inside. Keep shrinking. At 100 nm you approach the size of large viruses: influenza is roughly this scale, and many bacteriophages have heads in this ballpark, with tails that extend longer. At 10 nm you enter the territory of large protein complexes and the thickness of membranes. At 1 nm you are down to the diameter of a small protein's folded backbone or the width of DNA. Below that lie the spacing of atoms in a solid and the dimensions that make chemistry feel like geometry: typical interatomic distances in molecules are on the order of an angstrom, where $1\,\text{Å} = 0.1\,\text{nm}$.

The point of this ladder is not to impress with tiny numbers; it is to make a practical claim about what can and cannot be distinguished by different kinds of microscopes. The trouble begins around the level where cells hand off responsibility to molecules. A cell at $10\,\mu\text{m}$ can be seen. A bacterium at $1\,\mu\text{m}$ can be seen. A virus at $100\,\text{nm}$ is already becoming difficult, and a protein at $5\,\text{nm}$ is not merely small but, under ordinary visible light, effectively shapeless.

At this point it is tempting to think the obstacle is simply magnification. If something is too small, enlarge it more. That intuition is so natural that it has driven centuries of instrument-making. The first great microscopic revelations did, in a sense, come from magnification: Robert Hooke's *Micrographia* (1665) made the microscopic world famous by turning flea legs and plant tissues into elaborate engravings. Soon after, Antonie van Leeuwenhoek, working with tiny single-lens microscopes of exceptional craftsmanship, described what he called "animalcules" swimming in drops of water: bacteria and protozoa, though he did not have those words. The cultural shock of realizing that a drop of pond water contained a zoo was powered by the feeling that the eye had been extended, made stronger.

Yet magnification can be a kind of optical bluff. You can always make a blurry thing bigger. A low-resolution digital photo looks no sharper when zoomed; the pixels grow, but the information does not. Microscopes have their own version of pixelation, but it is not set by a grid of detectors. It is set by the wave nature of light itself. The critical distinction is between *making something look larger* and *resolving* detail.

Resolution is the ability to tell that two nearby features are separate. Imagine two streetlights seen from far away at night. If they are close together, your eye and the atmosphere smear their light into overlapping halos until they merge into one bright blob. You can magnify that blob with binoculars and it will become a larger blob. Only when your optics can collect and focus the light with enough fidelity do the two lamps split apart. In microscopy, the objects are not streetlights but tiny scattering centers in a specimen, and the halos are not merely imperfections but an unavoidable consequence of diffraction.

Diffraction is what happens when a wave encounters an aperture or an edge. Instead of travelling in perfectly straight rays, the wave spreads and produces an interference pattern. In a microscope, even a perfect lens cannot focus light from a point in the specimen into an infinitely small point in the image. The best it can do is a small, structured spot called the point spread function: a bright central disk with surrounding rings, like a target. If two points are close enough that these disks overlap strongly, they cannot be cleanly separated. The image begins to lie: two become one.

This limit was given a crisp theoretical form in the nineteenth century by Ernst Abbe, working with the Zeiss company at a moment when microscope-making was becoming a serious industrial science. Abbe realized that resolution is not just a matter of making better glass. It depends on the wavelength of light and on how wide an angle of that light the lens can gather. A convenient rule of thumb for the best possible resolution d of a conventional light microscope is

$$d \approx \frac{\lambda}{2\,\mathrm{NA}},$$

where λ is the wavelength of the light and NA is the numerical aperture, a measure (built from geometry and refractive index) of how much of the diffraction pattern you can capture. Visible light runs roughly from 400 nm (violet) to 700 nm (red). Even with very good lenses and immersion oils, NA for light microscopy is on the order of 1 to 1.4. Put those values together and you land at a best-case d of about 200 nm, sometimes a bit better under ideal conditions, but not by an order of magnitude.

That number, 200 nm, is not a casual engineering limit; it is the point where visible light begins to run out of teeth. It is also why so much cellular biology was historically a science of outlines and stains. Structures larger than a couple of hundred nanometers—nuclei, some organelles, whole bacteria—can be resolved directly. Smaller things are detectable only indirectly: as a blur, as a change in brightness, as a fluorescent dot, as motion, as the consequences of their activity. The microscope becomes less like a camera and more like a smoke detector: it can tell you something is happening, but it cannot show you the gears.

A protein is typically a few nanometers across. A ribosome, one of the cell's most important molecular machines, is about 20 to 30 nm in diameter. A membrane is about 5 nm thick. The DNA double helix is about 2 nm wide. These sizes sit not just below the 200 nm resolution of visible light microscopy, but often below it by factors of ten to a hundred. Under a conventional light microscope, proteins are not small versions of visible objects; they are below the resolving power of the instrument in the same way that individual grains of pollen are below the resolving power of your unaided eye at arm's length.

There is a second, subtler problem that arrives even before the diffraction limit becomes absolute: many biomolecules are not naturally high-contrast objects in visible light. A glass bead in water is hard to see because it bends light only slightly differently from the surrounding liquid. Many proteins are like that: mostly made of light elements—carbon, nitrogen, oxygen, hydrogen—and floating in water, with refractive indices that differ from the medium by a modest amount. The result is that even when a protein is present, it may not cast a bold optical shadow. Fluorescent labeling can make molecules *detectable* as points of light, and the last few decades have produced ingenious ways of using fluorescent blinking and computational tricks to beat classical resolution in special cases. But that kind of seeing is often about *where* a tagged molecule is, not *what shape* it takes, how it bends, what pockets it opens or closes, or how it fits with another molecule like a hand in a glove.

To feel the mismatch between biological scales and optical limits, it helps to keep one everyday comparison in mind. If a $10\,\mu$m cell were scaled up to the size of a sports stadium, roughly 100 m across, then a 5 nm protein would scale to about 5 cm: the size of a small plum or a ping-pong ball. In that stadium-city, proteins would be everywhere, countless machines doing their jobs. But if your "camera" could only resolve details larger than about 2 m—the scaled-up equivalent of 200 nm—then you could map the stadium's boundary and perhaps distinguish a few large structures, but every machine on the field would vanish into an unresolved texture. You might infer their presence from the stadium's function, or from colored beacons attached to them, but you would not see their design.

The predicament becomes clearer if you think of resolution as a kind of minimum spacing that the wave allows you to discriminate. For visible light, that spacing is measured in hundreds of nanometers. For molecules, the spacing that matters—the distance between one helix and another, between one side chain and another, between one ligand and its pocket—is measured in angstroms to a few nanometers. These are not just smaller by degrees; they belong to different regimes.

The phrase "beyond the rainbow" is literal here. The rainbow is the human-friendly span of wavelengths that our eyes can perceive. Shorter than violet is ultraviolet; longer than red is infrared. Both can be used for imaging in certain contexts, but neither solves the basic scaling problem for molecular structure. Ultraviolet light has a shorter wavelength than visible, so in principle it can resolve somewhat finer detail, but it is still measured in hundreds of nanometers down to a few tens of nanometers, and it brings new troubles: absorption, photochemistry, damage, and optics that are harder to make and use. Going to shorter wavelengths than that pushes you out of ordinary "light" entirely and into X-rays, which are indeed short enough to probe atomic structure. But X-ray imaging has its own challenges: focusing X-rays is difficult, and biological samples are again damaged by radiation. The history of structural biology is, in many ways, a history of hopping across these wavelength regimes and paying the price each time.

For a long period, microscopists tried to pay the price in glass. They worked to reduce aberrations, to improve polishing, to find better lens designs. These efforts mattered enormously. But Abbe's insight had an unsettling implication: even if you build a perfect microscope, diffraction still blurs a point into a spot whose size is tied to wavelength. The sharper image is not limited by craft alone; it is limited by physics.

That is why the term *diffraction limit* has such a particular emotional charge in biology. It sounds like a bureaucratic barrier, a line on a chart. In practice it has shaped what questions felt approachable. Under the light microscope, a dividing cell is dramatic: chromosomes condense, align, and separate. Neurons branch and grow. Embryos fold. These events unfold at scales where light can give you shape and motion. But the moment a question turns on the arrangement of a receptor's binding site, the symmetry of a viral shell, or the precise way an enzyme cradles a substrate, visible light becomes a storyteller who can only speak in silhouettes.

The nanometer world is not simply smaller; it is also faster and more crowded. Molecules are buffeted by thermal motion, colliding constantly. A protein does not sit like a fixed machine on a bench; it jitters, flexes, and samples shapes. Even the concept of a "picture" begins to change. At human scales, objects have stable boundaries. At molecular scales, the boundaries blur into probability clouds and dynamic conformations. Yet for all that motion, the details matter. A drug works because it fits. A mutation matters because it changes a surface by a fraction of a nanometer. A virus succeeds because its proteins present the right contours at the right spacing.

So the real obstacle is not that biomolecules are shy or hidden in a corner. It is that, under visible light, their details are fundamentally undersampled by the wave that is doing the looking. The world below a couple of hundred nanometers is a place where the ordinary microscope can point and say, "Something is here," but cannot, by its nature, draw the object's face.

When biology depends on faces, not just footprints, the only honest solution is to change the kind of wave you use for seeing. The nanometer is not merely a smaller unit; it is a boundary marker between what can be directly resolved with the rainbow's wavelengths and what requires a different kind of illumination altogether.

1.2 Waves with Teeth: Why Electrons Outperform Light

If visible light is a soft brush, electrons are a needle. Both can form images, and both can be persuaded into behaving like waves, but the scale of their ripples is radically different. The consequence is not subtle. A wave with long crests can only trace broad outlines; a wave with tightly spaced crests can, at least in principle, pick out the spacing of atoms.

The idea that electrons have a wavelength sounds like poetry until you remember how often everyday devices rely on waves that you cannot see. Radio is not a metaphor; it is electromagnetic waves filling a room, nudged into carrying a voice. X-rays are waves too, though too short for the eye and too energetic to be gentle. Electrons feel like particles because they arrive as localized hits on a detector, but they travel as something that spreads, bends, and interferes. This wave—particle duality is one of the central lessons of twentieth-century physics, and microscopy is one of its most practical uses. An electron microscope is not a glorified rifle firing pellets at a target; it is an instrument that shapes electron waves, lets them interact with matter, and then reconstructs an image from how those waves have been changed.

The decisive advantage begins with a single relationship proposed by Louis de Broglie in the 1920s: a moving particle has an associated wavelength. The faster the particle, the shorter the wavelength. For electrons accelerated through the voltages used in a transmission electron microscope (TEM), that wavelength is extraordinarily small. Visible light is typically around 500 nm, give or take. By contrast, electrons at the tens to hundreds of kilovolts used in TEM have wavelengths in the picometer range (10^{-12} m). That is not a modest improvement; it is a change of universe. Picometers are smaller than atoms. Even allowing for the practical imperfections of real instruments, the raw *potential* is there: electrons can carry information about features at the scale where chemistry happens.

At a dinner table, the cleanest way to express this is to replace "wavelength" with "ruler." Light arrives with a ruler whose tick marks are hundreds of nanometers apart. Electrons arrive with a ruler whose tick marks are thousands of times closer together. If you are trying to measure something the size of a protein, the light ruler is so coarse that the object falls between the ticks. With electrons, the ticks are close enough that the protein's bumps and grooves could, in principle, be measured.

That word *in principle* matters. Microscopists learned early that the wavelength does not automatically translate into sharpness. A finer ruler helps only if you can read it. The electron microscope has to create and manipulate an electron wave in a way that preserves the information it carries. That requires a whole environment engineered around the electron's needs, not ours.

A historical snapshot makes this vivid. In 1931, Ernst Ruska and Max Knoll built one of the first electron microscopes. Their motivation was straightforward: if resolution is limited by wavelength, then shorter wavelengths should resolve finer detail. Electron waves promised that. But the path from promise to picture was not just a matter of swapping a lamp for an electron source. Their early instrument was an act of radical reinvention. Glass lenses were replaced by electromagnetic coils that could bend electron trajectories. The entire beam path had to be kept free of air. The specimen had to be thin enough that electrons could pass through it. The resulting images were crude by modern standards, but they were a proof of concept: the laws that limited light microscopy were not laws against microscopy itself; they were laws about which wavelength you had chosen.

The basic architecture of a TEM still follows that early logic. Electrons are generated at a source and accelerated to high speed by an electric potential. They are shaped into a beam and focused by electromagnetic lenses, not made of glass but of magnetic fields. The beam passes through a specimen that must be extremely thin, and then the emerging electron wave is magnified and projected onto a detector. It is, in an abstract way, similar to an optical microscope. In practice it feels like another species of machine entirely.

The first reason is simple: electrons do not travel well in air. Air is a thick forest of molecules. For light, that forest is almost transparent. For electrons, it is a scattering carnival. An electron trying to fly through air at the energies used for imaging will collide, change direction, lose energy, and generate spurious signals. The beam will spread and degrade long before it reaches the specimen, and whatever emerges will be hopelessly scrambled. The solution is a vacuum: the column of an electron microscope is evacuated until there are so few gas molecules that electrons can travel meters without a meaningful collision.

Vacuum is not a small inconvenience; it is a profound constraint. Living cells cannot survive in it. Liquid water boils away. Many materials outgas, shedding molecules into the chamber and contaminating surfaces. Even fingerprints become troublesome. An electron microscope feels less like a piece of laboratory bench equipment and more like a spacecraft component: sealed, pumped down, stabilized, and constantly protected from the ordinary messiness of the world.

The second reason is that electrons are charged. That is what allows electromagnetic lenses to steer and focus them, but it also means the beam is exquisitely sensitive to stray fields. Small vibrations, electromagnetic noise from nearby equipment, instabilities in the high-voltage supply, drift in the lenses as they warm, all jostle the beam. A good optical microscope can be carried across a room and set down on a table. A high-end electron microscope is installed like an organ: anchored on vibration isolation, sheltered from temperature swings, and treated as a building within a building.

And then there is the lens itself. It is tempting to imagine that once you have a tiny wavelength, you can focus it to an equally tiny point. Real lenses are not perfect, whether they are glass or magnetic. In electron microscopy, the main culprit has long been aberrations, especially spherical aberration. In plain language, spherical aberration means that rays (or, more properly, wavefront components) passing through the outer parts of a lens are focused slightly differently from those passing near the center. Instead of bringing all contributions to the same point, the lens spreads them. The result is blur that cannot be removed just by turning a knob for better focus. Electron microscopes also face issues of partial coherence: the electron source is not a mathematically perfect point emitting perfectly uniform waves. The beam has a spread of energies and angles. That spread smears the finest interference patterns and reduces the contrast of the highest-resolution details.

Microscopists therefore speak not only about what the wavelength would allow, but about what the instrument can actually deliver: the *point resolution* and the *information limit.* These phrases sound abstract, but they describe an everyday experience. You can own a camera with a theoretical sensor resolution far higher than you ever see in your photographs, because the lens, the shake of your hands, and the haze of the air all conspire to throw away information. In electron microscopy, the throwaway happens through aberrations, instabilities, and coherence limits. The wavelength sets the upper ceiling; the microscope decides how far below that ceiling you will live.

The reward for managing all of this is a kind of seeing that light cannot match. Under the right conditions, electron microscopes reveal periodic lattices in crystals, the ordered spacing of atoms in thin materials, and the internal architecture of viruses and organelles at a clarity that would be miraculous under visible light. It is easy to say "electrons have shorter wavelengths" and leave it there. The more important point is that electrons do not merely show smaller things; they show a different *kind* of detail. At nanometer and angstrom scales, shape is inseparable from function. To see a protein's fold is to see where it can bind, what it can flex, and how it might be blocked or mimicked by a drug.

But electrons extract their payment in damage and in unnatural conditions. Even before discussing radiation damage in detail, it helps to notice a basic asymmetry: a visible-light microscope is a gentle observer by default. Photons in the visible range carry enough energy to excite certain molecules, which is why fluorescent dyes work and why intense light can bleach them. Yet for many specimens, ordinary illumination is compatible with life. You can watch cells crawl, divide, and respond to stimuli. You can keep them warm, feed them, and let them behave naturally.

An electron beam is never that kind of observer. The beam's electrons are energetic. They interact strongly with matter—strongly enough to be useful for imaging, but also strongly enough to cause changes. Even when the goal is to take the lightest possible look, an electron microscope is not a window onto living processes in real time. It is, more often, a camera that takes portraits under harsh studio lights in a place where the subject cannot remain alive.

This harshness begins with the requirement that specimens be thin. In a TEM, the electrons must pass through the sample. If the sample is too thick, multiple scattering events scramble the information: electrons ricochet, lose their original relationship to the structures they encountered, and the image becomes difficult to interpret. For materials science, one can cut ultrathin slices, polish, or thin samples with ion beams. For biology, thinness is an immediate problem because life is mostly water, and water is not cooperative in a vacuum.

The earliest biological electron microscopy solved this by removing water: by drying, fixing, staining with heavy metals, and embedding in plastics. Those methods produced images that were revolutionary and often beautiful, but they carried the suspicion of distortion. Drying can collapse delicate structures. Chemical fixation can lock molecules into unnatural poses. Stains add contrast by coating or filling spaces, which is helpful, but it is also a kind of interpretive act: you are not seeing the molecule directly so much as seeing where the stain has chosen to go.

Cryo-electron microscopy was born from the desire to keep biological structures closer to their native state—hydrated, unstained, and not flattened by dehydration—while still meeting the electron microscope's demands. That is why, when electrons are introduced as waves with tiny wavelengths, the story does not stop at triumph. The shorter ruler comes with a laboratory that is colder, emptier, and more hostile than any living thing would choose.

There is also a subtle conceptual cost. With light, we are used to thinking of images as formed by intensity: brighter means more light. Many biological specimens under visible light are nearly transparent, so techniques like phase-contrast and differential interference contrast were invented to turn tiny changes in how light is delayed and bent into visible contrast. Electron microscopy inherits the same problem in a more demanding form. Unstained biological specimens are made mostly of light atoms, and the most important information they imprint on an electron wave is often a change in phase—a tiny delay or advance in the wavefront—rather than a dramatic absorption that would show up as simple darkness.

Phase is the part of a wave that tells you where you are on its cycle: crest, trough, or something in between. Two waves with the same amplitude can add together or cancel depending on their relative phase. Biological material can shift the phase of the electron wave passing through it, and that phase shift encodes structure. The challenge is that detectors record intensity, not phase directly. The microscope must therefore be operated in a way that converts phase differences into measurable intensity differences. In practice that often involves deliberate defocus and careful control of the electron optics. The electron microscope, for all its apparent hardness, frequently works by listening to subtleties in interference.

This is another place where the wave description matters. If electrons were only particles, you might imagine simply counting how many arrive in each spot. But the richest structural information is not merely in how many electrons pass through; it is in how the specimen reshapes the electron wave. That reshaping is what allows, in favorable cases, angstrom-scale features to be recovered. It is also why the microscope's coherence and aberrations matter so intensely: interference patterns are fragile. Smear them, and the finest information vanishes even if the wavelength would have allowed it.

A concrete example helps anchor these abstractions. Consider a virus particle around 100 nm across. Visible light can detect something at that scale under special circumstances, especially if it is fluorescently labeled or if many particles cluster together. But the virus's defining features—the arrangement of its surface proteins, the symmetry of its capsid, the way a protruding spike tilts and flexes—live at the level of a few nanometers down to angstroms. Those are exactly the dimensions that decide how the virus attaches to a cell and how antibodies might block it. With electrons, the wave's short wavelength means those features can modulate the scattered wave in a way that is, at least, not forbidden by physics. With light, they are blurred together before they ever reach the camera.

Yet to exploit that advantage, the virus cannot be floating in a warm droplet of water on a glass slide. It must be placed in a thin film, inside vacuum, and protected against both evaporation and the violence of the beam. Even when the specimen is not a whole virus but a purified protein, the same logic applies. The very act of using electrons to see the nanometer world forces biology into an alien theater where the usual comforts—air, warmth, liquid water, life—are removed.

That tension is part of what makes cryo-EM feel, to practitioners, like a discipline of negotiation. The instrument offers extraordinary resolving power, but only if the sample is prepared in a way that is compatible with vacuum and beam, only if the optics are stable enough to preserve high-frequency information, and only if the weak, phase-based signals can be converted into contrast without destroying the structure they represent. The electron's short wavelength gives microscopy its teeth; everything else is learning how to bite without crushing what you want to study.

An electron microscope, then, is not simply a more powerful light microscope. It is a different treaty with nature. The reward is access to a world where the spacing between atoms is no longer an abstract number in a chemistry book but a measurable distance, and where biological machines can, with enough care, be seen close to the scale at which they do their work. The price is that the act of looking must be staged in vacuum, through electromagnetic lenses that are powerful but imperfect, using waves so sensitive that they demand quiet, stability, and cold.

1.3 Too Much Seeing Kills the Specimen

A familiar frustration in photography is the dim room: you raise the exposure, the image brightens, the noise recedes, and the scene looks more like what your eye imagined. In electron microscopy that impulse is dangerous. Turn up the exposure and the picture does improve—for a moment. Then the subject begins to change under the beam. The microscope is not only a camera; it is also a source of energetic projectiles, and the specimen is not only an object; it is a delicate arrangement of chemical bonds that can be rearranged by the act of observation.

Electron irradiation damages biological specimens through excitation/ionization and ensuing chemical reactions. Those words can sound remote until you translate them into ordinary chemistry. An energetic electron passing through a thin film of ice and protein can knock electrons off atoms (ionization) or lift them into higher-energy states (excitation). Either way, the local chemistry becomes unstable. Bonds that held a protein's shape begin to break. Reactive fragments form and attack nearby groups. Small atoms shift. Side chains that helped define a binding pocket lose their positions. If the specimen contains water—and biology always does—radiolysis products can appear, meaning the water itself is split into reactive species that diffuse and cause further damage. What a microscopist experiences as a fading of high-resolution detail is, at the molecular level, a slow collapse of the very structure the image is meant to capture.

One reason this matters so much for cryo-electron microscopy is that biological structure is information stored in arrangement. A ribosome is not just a pile of atoms; it is a particular pile. The same atoms in a different geometry would be useless. The atomic portrait one hopes to draw is therefore not simply a record of composition, but a record of *relationships*: this helix nestled against that beta sheet, this charged patch facing that phosphate backbone, this pocket shaped to accept that ligand. Radiation damage attacks relationships first. Long before the specimen looks visibly "burned," the fine correlations that encode function begin to wash out.

The earliest electron microscopists saw this as a practical nuisance, not an abstract limit. If you hold a beam on a sensitive biological sample, the image can drift in character: contrast changes, the sharpness softens, and in extreme cases the specimen develops obvious defects. Older practitioners sometimes describe a dramatic symptom called "bubbling," where irradiated regions in the frozen film appear to sprout bright voids, as if the sample were boiling. That is not a metaphor. Gases can be generated by radiation chemistry, and under the vacuum of the microscope they have nowhere to go. The effect is visually striking, but the real tragedy has already happened earlier, at doses too low to announce themselves so theatrically, when angstrom-scale order has begun to disappear.

This is where cryo-EM becomes a discipline of rationing. Every electron that passes through the specimen carries potential information, because it has interacted with the structure in some way. But every electron also adds to the damage. The result is an information bottleneck: you cannot simply increase exposure to get clearer images because the specimen changes (or is destroyed) before enough signal is gathered. In ordinary microscopy, the limiting factor is often the instrument. In biological electron microscopy, a major limiting factor is the specimen itself.

Microscopists talk about a *dose* the way doctors talk about radiation dose: the amount of irradiation delivered to a given area. In cryo-EM it is often expressed as electrons per square angstrom, a unit that sounds arcane but matches the scale of the structures at stake. The intuitive idea is simpler. Imagine paying for an image with a currency that is also corrosive. Each coin buys a little clarity but eats away a little of the subject. Spend too little, and the image is dominated by noise. Spend too much, and the structure that would have made the image meaningful has been altered. The art is to spend just enough to extract the information you need while leaving the specimen sufficiently intact during the exposure.

That trade-off is why low-dose imaging feels, in practice, like working in the dark. A novice might expect that a modern electron microscope, with its imposing column and sophisticated controls, would produce high-contrast, crisp pictures on demand. Instead, the raw images of unstained biological specimens can look like faint ghosts in granular fog. The fog is not merely poor technique; it is the consequence of ethical restraint. You are refusing to brighten the scene because brightness comes at the cost of truth.

The low-dose mindset also reorganizes the workflow at the microscope. In a light microscope, you focus on the very object you intend to photograph. You adjust and refine, watching the specimen as long as you like. In cryo-EM, that casual attention would be fatal. If you focus and align on the target, you are spending your precious dose budget before the actual image is recorded. So the microscope is operated with a kind of misdirection. The operator focuses on an adjacent area, aligns the beam and optics away from the region of interest, and only then briefly exposes the chosen spot for acquisition. The main exposure is kept short and carefully measured, not because the operator is impatient, but because the specimen has an extremely limited tolerance for scrutiny.

The idea that there is a *critical dose*—a dose beyond which useful structural information decays sharply—was articulated and explored in foundational work by Robert Glaeser (1978), who helped clarify how radiation damage in biological electron microscopy behaves and why cooling helps but does not cure it. The phrase "critical dose" is not a magic threshold where everything suddenly fails; it is more like a cliff edge approached gradually. Low-resolution features may survive longer, while high-resolution details fade quickly. That uneven fading makes intuitive sense. Broad shapes depend on many atoms moving coherently; small displacements can leave the silhouette recognizable. Atomic-level detail depends on precise positions; small displacements erase it.

Cooling to cryogenic temperatures changes the damage progression and historically improved diffraction lifetimes; nevertheless, radiation damage remains a fundamental limit in cryo-EM, forcing low-dose imaging strategies. Again, the physics can be made concrete. At room temperature, radiation-generated radicals and fragments can move around readily. Chemical reactions proceed quickly. Cooling slows diffusion and many reaction pathways, effectively giving the structure a better chance of holding still long enough to be recorded. Freezing also immobilizes the sample mechanically, reducing motion that would blur the image even if the structure were chemically intact. But cooling does not stop the initial ionization and excitation events. The first insult still happens. Cryogenic temperatures are not a shield; they are a way of keeping the consequences of damage from spreading quite as fast.

There is a second, almost philosophical reason freezing is indispensable: it allows the specimen to remain hydrated in vacuum. Water is life's solvent, but in the microscope's vacuum it would evaporate instantly. The solution is to lock water into a solid form. Ordinary freezing, however, produces ice crystals, and crystals are disruptive. As they form, they exclude solutes, concentrate salts, and can physically shove and tear delicate structures. Their ordered lattice also strongly diffracts electrons, adding confusing patterns that overwhelm the subtle signal from a protein. Cryo-EM therefore relies on a particular kind of frozen water: vitreous ice, sometimes called "glass ice." This is water that has been cooled so rapidly that it does not have time to crystallize; it becomes an amorphous solid, structurally more like a frozen snapshot of the liquid than like a snowflake.

The practical act is almost theatrical. A tiny droplet of sample is applied to a grid—a delicate metal mesh supporting a thin film—and most of it is blotted away, leaving a film that is only tens to a few hundreds of nanometers thick. Then the grid is plunged into a cryogen so cold and so fast at drawing heat away that the water vitrifies. The specimen is trapped in place, surrounded by a glassy matrix that preserves its shape and keeps it in a hydrated-like environment while the microscope does its work in vacuum. The result is a biological specimen presented in a state that is neither living nor chemically fixed in the traditional sense: it is immobilized in a way that, at its best, preserves native structure without the heavy-metal disguises of earlier methods.

That word "best" matters because vitrification is itself a form of compromise. The frozen film must be thin enough for electrons to pass through with limited scattering, but thick enough to accommodate the molecules and keep them away from surfaces that could denature them. The ice must be glassy, not crystalline. The molecules must be well distributed and not stuck in preferred orientations. And through all of this, the sample must tolerate being placed into vacuum and held at cryogenic temperature without acquiring contamination. The frozen grid is a delicate object: a few millimeters across, but carrying within it a landscape of nanometer-scale specimens that will be exposed to a beam capable of breaking bonds.

From the outside, cryo-EM is often portrayed as a triumph of technology: brighter sources, better lenses, better detectors. From the inside, it can feel like an art of restraint. The most valuable skill is not pushing the microscope to maximum power, but choosing which electrons to spend and where. The dose budget is not an accounting trick; it shapes every decision. You choose a region of thin, clean ice because extra thickness demands extra dose. You avoid unnecessary exposures because every glance costs. You accept that a single image will be noisy because you have decided not to destroy the specimen for the sake of cosmetic clarity.

The dose-limited nature of cryo-EM also changes what it means to trust an image. In many forms of microscopy, the limiting factor is interpretive: the image is clear, but what does it mean? In cryo-EM, the image may be faithful but faint. The graininess is not a sign of carelessness; it is a sign that the microscopist has not bullied the specimen into confessing. The data arrive as whispers because the specimen cannot survive shouting.

A useful comparison is with listening to a quiet radio station far from the transmitter. You can increase the volume, but you also increase the hiss. In cryo-EM, increasing dose increases the volume of signal, but it also changes the station itself. You are not merely amplifying; you are burning the antenna as you listen. That is why the idea of "information per unit damage" became so central in thinking about the limits of biological electron microscopy. The microscope may be able to resolve angstroms, but the specimen may not be able to *stay itself* long enough to deliver angstrom information.

This constraint is also why structural biology has always been, in part, a science of cleverness under pressure. When X-ray crystallography delivers atomic models, it does so by spreading the dose across an enormous number of identical molecules in a crystal, sharing the damage among them. Cryo-EM's single-particle approach, when it works, achieves a similar sharing in a different way: not by locking molecules into a crystal lattice, but by imaging many individual copies, each one sacrificed a little to contribute a little. The dose on any one molecule is limited, but the dataset as a whole can be rich. Even before the computational details enter, the logic of survival has already forced a statistical strategy.

There is a human side to this as well. Standing at a high-end electron microscope, an operator can feel an odd mix of power and fragility. Power, because the machine can, under the right circumstances, reveal atomic-scale order. Fragility, because the specimen—a few proteins in a thin film of glassy ice—is so easily ruined by a moment's carelessness: too much exposure while focusing, an unnoticed patch of crystalline ice, contamination creeping in as frost, a slight drift that smears the already scarce information. The operator's hands are on controls that shape electromagnetic fields and high voltages, but the success of the experiment can still hinge on something as humble as the cleanliness of a grid or the steadiness of a temperature.

Cryogenic protection, then, is not an optional flourish. It is the practical response to a harsh truth: the act of seeing at the nanometer and angstrom scales can be inherently destructive. Cooling gives molecules a fighting chance, and careful low-dose practice stretches the narrow time window during which structure is still structure. For a brief exposure, the protein's bonds hold, its atoms remain near their places, and the electron wave that passes through carries away an imprint that can be interpreted later. The image is bought at the price of change, and the craft lies in keeping that price just low enough that the portrait still resembles its subject.

1.4 Whispers in the Noise: How Invisible Matter Becomes an Image

A first encounter with a raw cryo-EM micrograph can be oddly deflating. After the vacuum, the cold, the careful low-dose discipline, the image on the screen may look like little more than grey grit. Somewhere in that grit are the molecules you came to see, each one only a few nanometers across, each one imaged with so few electrons that the picture seems to apologize for itself. The temptation is to assume something has gone wrong: poor sample, poor microscope, poor operator. Often the opposite is true. The faintness is the signature of restraint. The microscope has been operated gently enough that the specimen has not been obliterated, which means the data arrive as whispers.

Why are biological molecules such poor subjects even for an electron beam? Part of the answer is simply that they are built from light atoms. A gold nanoparticle is an electron microscope's dream: heavy nuclei, strong scattering, high contrast. A protein is made mostly of carbon, nitrogen, oxygen, and hydrogen, and it sits embedded in a thin layer of vitreous ice made of the same light elements. Protein and ice differ only slightly in how strongly they disturb the passing electron wave. The specimen is real, but its imprint is subtle.

A second part of the answer is the dose budget. If one could illuminate a protein with an arbitrarily intense beam, the weak imprint could be compensated by sheer numbers of electrons, averaging out noise by brute force. But electrons are not polite, and the dose that a specimen can survive is limited. Single cryo-EM images of unstained biological particles are extremely noisy because dose must be low; extracting structure becomes a statistical problem solved by alignment and averaging across many particle images (and/or many frames). The practical consequence is that cryo-EM rarely succeeds by taking one perfect picture. It succeeds by taking many imperfect ones and letting patterns accumulate.

To understand how those patterns can be recovered, it helps to step back from the language of "brightness" and "darkness" and think again in waves. In an optical photograph, much of what you see is intensity contrast: a dark object absorbs or blocks light, so fewer photons reach the camera. Unstained proteins do not absorb electrons in that simple way. Instead, they act more like a piece of glass in water: they delay the wavefront slightly, bending it and shifting its phase. Phase is the bookkeeping of a wave's timing—where each part of the wave sits in its cycle. Change the phase and you have changed the wave, even if the overall intensity is nearly the same.

The trouble is that detectors record intensity. They count arrivals. They do not directly report phase. If an electron wave emerges from a specimen with altered phase but nearly unchanged intensity, the detector would, in the most naive arrangement, see almost nothing. Cryo-EM therefore relies on a trick that has a long history in microscopy: phase contrast. The basic idea is to convert phase differences into intensity differences by letting the wave interfere with itself in a controlled way. When different parts of a wave are out of step, they can add or cancel, producing detectable bright and dark variations. The specimen's structure is then translated into contrast, not because it is opaque, but because it has reshaped the wavefront.

In practice, one of the most common tools for creating phase contrast in cryo-EM is deliberate defocus. Defocus sounds like a mistake, and in everyday photography it is. In electron microscopy it can be a choice. If the microscope were set to perfect focus for a weakly scattering, unstained biological specimen, the image could be so low in contrast as to be nearly unusable. By defocusing—by setting the lenses so that the image plane is slightly above or below the specimen plane—the microscope turns subtle phase shifts into more visible intensity modulations. Edges and features that were invisible at perfect focus gain contrast, allowing particles to be found and aligned.

This choice, however, comes with consequences. Defocus does not merely increase contrast; it also reshapes which details are faithfully transferred from the specimen to the image. The microscope does not pass all spatial scales equally well. Fine details correspond to rapid variations in the wavefront; coarse shapes correspond to slow variations. The microscope's optics, along with defocus and aberrations, modulate these spatial frequencies in a characteristic way. That modulation is described by a central concept in cryo-EM: the contrast transfer function, usually abbreviated as the CTF.

The contrast transfer function (CTF) describes how defocus and aberrations modulate Fourier components of the image; CTF oscillations mean some spatial frequencies are attenuated or inverted, so correction/estimation is necessary for faithful reconstructions. The words "Fourier components" can sound like a mathematical detour, but the idea is intuitive. Any image can be thought of as a mixture of patterns at different scales: broad gradients, medium-sized features, fine ripples. The CTF is the microscope's fingerprint on that mixture. Some scales are emphasized, some are damped, and some are flipped in sign, meaning that a feature that should appear bright may appear dark and vice versa. Defocus, then, is not free contrast; it is contrast with a price tag attached: a complicated, oscillating distortion that must later be accounted for.

How can one know what distortion a particular micrograph has suffered? Here cryo-EM reveals one of its strange beauties: even the noise contains information. When a micrograph is transformed into a power spectrum—a kind of map of how much pattern exists at different spatial scales—it often shows a series of faint, concentric rings. These rings are not features of the specimen. They are a record of the microscope's transfer behavior. Thon rings in the power spectrum provide a practical route to estimating defocus/astigmatism and thus characterizing the CTF for each micrograph. The name honors Fritz Thon, who described how such rings arise. They are, in effect, calibration marks that appear because the microscope has imposed a structured modulation on the image. Once that modulation is recognized, its parameters can be estimated and used to correct the data computationally.

Astigmatism is worth a brief aside because it is an everyday lens flaw in a more exotic setting. In an astigmatic lens, the focus differs depending on direction, rather like a slightly warped eyeglass lens that makes vertical lines and horizontal lines come into focus at different settings. In a TEM, small asymmetries in the electromagnetic lenses can create astigmatism, turning what should be circularly symmetric transfer into something slightly elliptical. Thon rings, when present, reveal this too: rings become ovals. What might have been dismissed as yet more noise becomes a diagnostic signal, allowing the operator and the software to infer how the microscope behaved during the exposure.

This mixture of physics and inference is why cryo-EM is sometimes described as a computational form of microscopy. The detector records an image, but the image is not yet a straightforward portrait. It is a portrait filtered through the CTF, blurred by motion, and corrupted by noise that arises because the dose was kept low enough to preserve structure. The way forward is to take many such images and let statistics do what dose cannot.

A concrete historical turning point here is the development of image averaging methods for biological EM, associated particularly with Joachim Frank's work from the 1970s onward and synthesized in his later writing (for example, his 2006 treatment of two-dimensional averaging techniques). The key insight is simple to state: if you have many noisy images of the same object, and if the noise is largely random, then averaging aligned images will reinforce the true signal and suppress the noise. The hard part is alignment. Molecules appear in random orientations and positions, and each individual image is so noisy that it may be difficult to see the object well enough to line it up by eye. Frank and others developed computational approaches that use cross-correlation and classification to align and group particle images, producing class averages where the molecule's features begin to emerge with clarity.

The logic resembles a familiar social experience. In a crowded room you might hear only fragments of a quiet conversation. If you could record a hundred similar conversations and synchronize them, the shared words would rise above the babble. Cryo-EM does something analogous. Each particle image is a noisy projection of a molecule. By finding many instances of similar views and averaging them, the shared structure becomes visible. The signal-to-noise ratio improves roughly with the square root of the number of independent images: four times as many particles gives about twice the clarity, all else equal. That is slow progress, which is why cryo-EM datasets often contain tens or hundreds of thousands of particle images, and sometimes more.

A complication immediately appears: molecules are not always identical copies. Proteins breathe; complexes adopt different conformations; some particles are damaged, partially assembled, or simply not the molecule of interest. Averaging indiscriminately would blur real biological differences into a mush. That is why classification matters. By grouping particles into classes that represent distinct views or distinct conformations, one can average within each class and preserve meaningful variation. Cryo-EM is therefore not just about gathering large numbers; it is about choosing which numbers belong together.

While averaging across particles is crucial, averaging can also occur within a single exposure, thanks to improvements in detectors. Traditional electron detectors often relied on a scintillator: electrons hit a material that emits light, and the light is then recorded by a camera sensor. This conversion adds blur and noise. In cryo-EM, where the specimen sets a hard limit on dose, wasted signal is painful. Detector performance (DQE/MTF) strongly impacts how much usable SNR can be extracted under dose constraints; direct electron detectors improved low-dose image quality and enabled practices (e.g., dose fractionation into movie frames) that support motion correction and better high-resolution transfer.

Two pieces of jargon are doing real work in that sentence. DQE is detective quantum efficiency: a measure of how well a detector preserves the signal-to-noise ratio of what hits it. A perfect detector would add no extra noise beyond the inherent randomness of counting electrons. Real detectors add noise and blur. MTF is the modulation transfer function: a measure of how well contrast at different spatial frequencies is preserved, essentially how much fine detail is smeared out by the detector. Direct electron detectors, as their name suggests, detect electrons directly rather than converting them to light first, improving both DQE and MTF in ways that matter enormously for low-dose imaging.

The practical revolution this enabled is dose fractionation. Instead of taking a single exposure of, say, a second, one can record a rapid series of frames during that second—a movie. The total dose is the same, but it is spread across frames. This turns out to be a gift because specimens move. Under the beam, even a vitrified sample can undergo tiny motions: the ice can flex, charge can build up and relax, the specimen can drift. In a single long exposure those motions blur the image, erasing high-resolution information. In a movie, the motion can be estimated by comparing frames, and then corrected computationally by aligning and summing them. The idea is again an act of rescue: information that would have been smeared into invisibility is recovered not by increasing dose, but by distributing the dose in time and using computation to undo the damage of motion.

That move—spending the same dose more wisely—is characteristic of the whole field. Because the specimen sets the dose ceiling, progress often comes from reducing waste: better detectors that throw away fewer electrons' worth of information, better optics and stability that preserve coherence, better algorithms that extract signal from noisy images without inventing structure that is not there. Even the microscope alignment practices reflect this thrift. One learns to focus and stigmated-correct off-target, to measure the CTF from the micrograph itself, and to let software correct what would be impractical to correct perfectly in hardware.

It is worth lingering on the emotional experience of this process, because it is not the usual romance of seeing. The raw micrograph does not flatter. The molecule does not pose. The picture does not arrive with a satisfying click. Instead, the structure appears gradually, like an image emerging in a developer tray in an old photographic darkroom. That analogy is imperfect—the "developer" here is mathematical rather than chemical—but it captures the peculiar suspense of cryo-EM. At first you have only suspicion: faint particles, barely there. Then, after alignment and averaging, recognizable shapes begin to cohere: the outline of a complex, a groove, a central channel. Finally, with enough data and careful correction of the CTF, finer features appear: helices as rods, beta sheets as ribbons, side-chain bumps at the highest resolutions. The molecule's identity moves from rumor to portrait.

A natural worry follows: if so much is computational, how does one avoid imagining what one hopes to see? This worry is healthy, and cryo-EM has built a culture around it. The CTF, for instance, is not a hand-waving correction; it is a physically grounded model whose parameters can be estimated from Thon rings. Averaging does not conjure structure out of nothing; it suppresses uncorrelated noise while preserving correlated signal. When methods are used responsibly, the computation is not decoration layered onto an image; it is the completion of an imaging system whose raw output is necessarily incomplete because the specimen could not survive a more forceful interrogation.

The microscope, the specimen, and the computer therefore form a single instrument. The beam provides the probing wave. The cryogenic preparation preserves structure long enough to be sampled. The optics impose a transfer function that can be characterized. The detector records the result with some efficiency. The computer then carries out the final act: estimating, aligning, correcting, and averaging, so that the faint phase imprints left by a molecule in glassy ice can be turned into a map of its shape.

The result is a new kind of realism. It is not the realism of a snapshot taken in bright light, but the realism of a coherent story assembled from many cautious glances, each one too gentle to be decisive on its own. In a field where too much seeing kills the specimen, the most faithful portrait is often the one built from the least invasive looks, combined with patience and an honest accounting of how waves, lenses, and noise have edited what the detector could record.

Chapter 2

Water That Refuses to Crystallize

At the heart of cryo-EM lies a paradox: to see life clearly, scientists must freeze it so quickly that water never has time to become ordinary ice. What sounds like a technical footnote turns out to be the chapter's drama, as tiny decisions about film thickness, metal supports, and the sample's uneasy response to electrons determine whether a molecular portrait emerges sharp or dissolves into blur.

2.1 When Water Becomes Glass

Freezing feels like one of nature's most straightforward tricks. Put a puddle outside on a cold night and by morning it has become something solid, something you can scrape from a windshield. In the kitchen, the change announces itself with familiar signs: the gloss of a liquid dulls, edges sharpen into facets, and—if you look closely—crystals begin to claim the water as their own. We are used to thinking of ice as *the* solid form of water, as if "frozen" automatically means "crystalline."

For people trying to photograph life's smallest machinery, that everyday intuition turns into a trap. Ordinary ice is not a helpful preserving blanket. It is an active reorganizer. It grows, pushes, excludes, concentrates, and scatters. It is also, in a very literal optical sense, loud: it produces strong, structured patterns that can overwhelm the faint traces from the biological matter you care about. The great conceptual step behind modern cryo-EM was not a new lens or a clever detector; it was learning how to make water solid without letting it do the things ice usually does.

The enemy is not cold. The enemy is the *lattice*.

When water crystallizes, it does not simply stop moving; it chooses a geometry. Molecules of H_2O arrange themselves into a repeating pattern—one that can extend for millions of molecular spacings if you let it. That repetition is what makes a crystal a crystal, and it is why snowflakes have their precise sixfold symmetry. The lattice is beautiful, but in the context of imaging it is disastrous for two different reasons.

The first is mechanical and chemical, the way freezing changes what is left behind. Ice crystals are not made of "water plus whatever else happened to be dissolved." The lattice strongly prefers pure water. As a crystal advances, it incorporates H_2O molecules and tends to reject many solutes—salts, sugars, buffer components, and, crucially, the biomolecules you are trying to observe. Those rejected ingredients get squeezed into the shrinking liquid regions between growing crystals. Concentrations rise locally, sometimes to levels that would never occur in a cell. pH can shift. Proteins that were comfortably dissolved can be forced together or partially unfolded. Membranes can be bent and torn as crystals expand and press. Even if the biological structure survives chemically, its position may not: crystals can shove particles into clumps or toward interfaces, leaving large regions empty.

The second problem is physical in a different sense: crystalline ice is a superb *diffractor*. A repeating lattice acts like a three-dimensional grating. When a beam of electrons passes through, the crystal's regularity produces strong, sharp diffraction spots and rings—structured signals that are easy to detect precisely because the crystal is so orderly. Biological material is the opposite. Proteins and nucleic acids are complicated, irregular objects; their signal is subtle, and it fades quickly in the noisy conditions imposed by radiation damage. If crystalline ice is present, it can dominate the image with patterns that have nothing to do with the specimen's shape. The background becomes organized and intense at exactly the spatial frequencies where you would like to measure the finest details of a macromolecule. It is like trying to hear a whisper while a metronome clicks loudly at many pitches at once.

These two failures—crystals that rearrange the sample and crystals that scatter strongly—are tightly linked to what makes ice familiar. The very property that gives a solid its facets and its strength, the repeating order, is what a structural biologist cannot tolerate.

So what would it mean to freeze water *without* letting it crystallize?

Water has another option, one that sounds like a contradiction: it can become a glass. Glass, in everyday life, is a solid that looks like a liquid arrested mid-motion. Window glass is not crystalline. Its atoms are disordered; they do not repeat in a tidy lattice. Yet it is rigid enough to tap with a fingernail. Water can do the same, forming what is often called vitreous ice—"vitreous" meaning glassy. The molecules stop moving, but they do not have time to line up into a repeating pattern. The result is a solid that is mechanically supportive and, crucially for imaging, lacks the long-range order that creates strong diffraction signatures.

The price of this trick is speed. Crystallization is not a moral failing of water; it is a kinetic race. To crystallize, molecules must first form a tiny embryo of the lattice—a nucleus—and then that nucleus must grow. Both steps take time. If cooling is slow, water molecules have enough mobility to explore configurations and settle into the energetically favored crystal. If cooling is breathtakingly fast, mobility vanishes before the molecules can organize. The liquid becomes a solid snapshot of itself, like a photograph taken with an impossibly short exposure.

That "snapshot" idea is more than a metaphor. In a hydrated biological sample, water is not a passive background; it is part of the structure. Proteins breathe—side chains shift, loops sample positions, small domains rock minutely. At room temperature the motions are continuous. Rapid freezing does not select one perfect pose; it captures an ensemble, each molecule caught in one of the conformations it happened to occupy. In practice, that is a gift. It means the specimen is preserved close to how it existed in solution, rather than being forced into the unnatural states produced by slow freezing and concentrated solutes.

The constraint is that water is very good at crystallizing. Pure water, given a leisurely descent in temperature, will almost always find its lattice. To outrun it, the cooling must be intense and the layer must be thin. Thickness matters because heat has to leave the specimen. A thick droplet is a thermal fortress: its interior cools slowly because heat must conduct outward through already-cooling material. A very thin film, by contrast, has a short path for heat to escape. When it is plunged into a cryogen, the temperature of the film can crash fast enough that nucleation and growth are stifled across most of the volume.

This is why the seemingly mundane detail of specimen thickness became central to the field's history. The transformation from "frozen sample" to "usable sample" depended on making layers of water so thin that they could be vitrified reliably, and on cooling them so rapidly that the molecules had no chance to arrange themselves into a crystal.

A useful way to picture the difference is to imagine embedding a small object—say a pebble—in two kinds of solidifying material. In crystalline ice, the advancing lattice is like a regimented construction crew laying bricks in a repeating pattern. Anything that does not fit the pattern is pushed aside or squeezed into the remaining gaps. In vitreous ice, there is no brickwork. The liquid simply stiffens everywhere at once, and the object stays where it was, surrounded by a disordered but solid matrix. The water still touches the biomolecule, still provides support, but it is no longer an architect with opinions.

That conceptual leap was made practical in the early 1980s, when Jacques Dubochet and his colleagues showed that thin layers of water could be vitrified in a way suitable for imaging hydrated biological specimens in the vacuum of a microscope. Their 1981 work with Alasdair McDowall is often remembered for its crispness: it turned a vague aspiration—"could we freeze without crystals?"—into a concrete method with criteria you could test. One could look at the frozen film and tell, by its diffraction signature, whether it was truly vitreous or contaminated with crystalline ice. The advance was as much about *diagnosis* as about technique. It gave the field a way to distinguish "cold" from "correct," to stop arguing by intuition and start verifying the water's physical state.

That ability to diagnose matters because vitrification is not a permanent moral victory. The glassy state is metastable. Warm it too much, even briefly, and it can devitrify—transform into crystalline ice. In practice this means that the entire handling chain has to respect water's impatience to become orderly. Samples must stay well below the temperatures at which ice can reorganize during transfer and loading. A moment of carelessness can turn a good grid into one full of crystalline patches, the microscopic equivalent of a fine varnish blooming into visible cracks.

It can be tempting, especially when one first hears about "glassy water," to imagine a perfect, featureless block of transparent solid in which molecules hang like fruit in clear jelly. Real vitreous ice is not quite that idyllic. It can contain stresses from rapid cooling, tiny thickness gradients, and regions near surfaces where molecules behave differently. Yet compared with crystalline ice, it is astonishingly kind. Its disorder means it produces a relatively smooth background. It does not generate the sharp diffraction rings that announce crystals. Its mechanical behavior, while delicate, is less likely to bulldoze macromolecules into unnatural concentrations during freezing.

One subtle point is worth lingering on because it overturns another everyday expectation. In ordinary life, freezing is often thought of as a way to *stabilize* things: bacteria in a freezer, cells in liquid nitrogen, a sample preserved for later. In cryo-EM, the goal is not merely to stop time; it is to stop time *without introducing a new structure of your own.* Crystalline ice is a structure—a strong one—with its own geometry and its own scattering power. Vitrified ice is as close as you can get to "no structure" while still being solid. The disordered arrangement is not a nuisance to be tolerated; it is the very reason the specimen remains interpretable.

The distinction also helps explain why the word "native" carries so much weight in this field. A protein in a cell is hydrated, surrounded by water molecules that form a dynamic, shifting shell. Remove the water and many proteins collapse or rearrange; the vacuum inside the microscope demands dryness, yet the biology demands hydration. Vitreous ice is a compromise engineered from physics: a way to keep water present while removing mobility. It is not the same as liquid water, but it preserves a critical aspect of the native environment—hydration—without allowing the sample to be torn apart by crystals or by the chemical concentration effects of slow freezing.

This is the point where the physics of water meets the mathematics of reconstruction. A cryo-EM image is a projection of a molecule's internal structure, but the molecule's contribution to the image is weak. Every unwanted scattering source—every patch of crystalline ice, every thick region of frozen buffer—eats away at the faint signal you need. Vitreous ice, being disordered, tends to add a more uniform haze rather than a set of strong, patterned features. Uniform haze can be averaged down and modeled; sharp diffraction features from crystals stamp their authority on the data in a way that is hard to ignore. Put bluntly, vitreous ice is quiet enough that the molecule can be heard.

The quietness has another advantage. Crystals do not merely diffract; they also introduce spatially varying contrast that can trick both human eyes and algorithms. In a micrograph with crystalline contamination, high-contrast ripples and rings may look like "structure" at first glance. Early practitioners learned, often the hard way, that a beautiful image could be a beautiful lie: what looked like a repeating biological pattern might be nothing more than the signature of ice. The discipline of looking for the telltale signs of crystals—either in diffraction patterns or in the texture of the image—became part of the craft. It is an acquired skill, closer to reading weather in the sky than to following a recipe.

If all of this makes vitrification sound like a narrow trick, it may help to remember how counterintuitive it was. For decades, the assumption was that water and high vacuum were fundamentally incompatible: to put something in the microscope, you had to dry it, stain it, fix it, coat it—transform it into something else that could survive the environment. The glassy-water approach said, in effect: keep the water, but deny it the chance to become ice. It is a sleight of hand performed not with mirrors but with kinetics.

The wonder is that the trick works at all. Water molecules are small, fast, and eager to find order. Yet with a thin enough film and a violent enough cooling rate, the liquid can be trapped in a solid state that still "remembers" the randomness of the liquid. Proteins and other macromolecules are caught where they were, surrounded by their hydration shells, not squeezed into briny pockets between expanding crystal fronts. The frozen water does not announce itself with a crystalline chorus of diffraction spots. It becomes, as much as physics allows, an invisible support.

There is a final, almost philosophical satisfaction in this. Life depends on water's unusual properties—its polarity, its hydrogen bonds, its ability to dissolve salts and shape biomolecular surfaces. In cryo-EM, those same hydrogen bonds would happily build an ice lattice that ruins everything. The method succeeds by exploiting a different oddity: water can be made into a glass, a solid that refuses the neatness of a crystal when time is taken away. The molecules of water do not stop being water; they simply lose the opportunity to become symmetrical.

And so the most familiar substance in biology is asked to do something profoundly unfamiliar: to freeze while staying disordered, to become solid without becoming crystalline, to hold the shapes of proteins steady without imposing a shape of its own.

2.2 A Grid, a Whisper-Thin Film, and a Fall Into Cold

The oddest thing about many near-atomic cryo-EM structures is how ordinary their starting point can look: a small tube of clear liquid and a square of metal the size of a fingernail. The sophistication hides in scale. A method that depends on water behaving like glass also depends on persuading that water to form, for a brief moment, a film so thin that it barely deserves the name "droplet," spread across a landscape of holes, edges, and supports. The film must then be cooled fast enough to outrun crystallization, and it must do so while holding fragile molecules in place, hydrated and intact, in a vacuum-bound future.

A grid is, at first glance, a modest object. It is usually a small disc of metal mesh—copper is common, and gold is increasingly popular—designed to fit into a specimen holder. The mesh provides mechanical strength, like the ribs of an umbrella. Across the squares of that mesh sits a much thinner supporting layer, the foil or film that actually carries the water. Many standard supports are holey carbon: an amorphous carbon film punctured by a regular array of micrometre-scale holes. Manufacturers often cite carbon thicknesses around 10–12 nm for common holey films: thin enough to be almost invisible to the beam, thick enough to handle.

Those holes are not a decorative pattern. They solve a conflict. If you place your sample on a continuous film, the underlying material contributes background scattering and can obscure faint details. If you remove the film entirely, there is nowhere for the sample to form a stable, thin layer; surface tension will pull a liquid into beads, and most proteins will end up somewhere inconvenient. A perforated film provides "windows" where molecules can be suspended in ice over open space, with enough surrounding support to make preparation feasible. It is a compromise that works surprisingly well, provided the liquid layer is the right thickness.

Thickness is the hidden master variable because it controls almost everything you care about: cooling rate, contrast, and the survival of the molecules at interfaces. Too thick, and the film cools more slowly and may crystallize. Too thick, and even if it vitrifies, the added water dilutes contrast: the beam passes through more material, and the molecule's signal is submerged in background. Too thin, and there may not be enough volume for the particles to sit comfortably; large complexes can be squashed against surfaces or excluded from the thinnest regions. Too thin, and a different hazard rises to the surface—literally.

A protein in solution is happiest when it is surrounded by water, not pressed against air. Yet a film spanning a hole has two major interfaces: the air–water surface on top, and the water–support surface (or another air–water surface) below, depending on the geometry. The air–water interface is a notoriously unfriendly place for many proteins. Some proteins adsorb to it within milliseconds, unfold partially, or adopt preferred orientations. Even when they do not denature, they can end up lining the surface like driftwood along a shore, all facing the same way. That makes later reconstruction harder, because you do not get the diversity of views needed to build a reliable three-dimensional map. Thinness helps vitrification, but it also increases the fraction of molecules that can reach an interface quickly. The very step that makes the water "good" can make the biology misbehave.

The practical choreography begins with a tiny volume of sample—often only a few microlitres—deposited onto the grid. At that moment, the grid is a wet object in a warm room, and time becomes an adversary. Water begins evaporating. Concentrations creep upward. Proteins begin diffusing to surfaces. A good preparation aims to minimize the interval between deposition and freezing, yet also to shape the liquid into a film of the right thickness across many holes.

That shaping is done by blotting: bringing filter paper into contact with the grid to wick away most of the liquid. Blotting looks crude, even faintly comic, given the precision expected of a technique that can place atoms. A piece of paper touches a metal mesh and the future of a dataset is decided. Yet blotting is also delicate. A fraction of a second longer can thin the film too far; a fraction shorter can leave it too thick. The pressure, humidity, and even the condition of the blotting paper influence how uniformly water is removed. Some laboratories treat the choice of paper and the adjustment of blotting force with the seriousness of a chef choosing a knife: the tool is simple, but the outcome depends on texture and touch.

As the paper pulls water away, the liquid that remains is sculpted by capillary forces. It clings to grid bars, pools near edges, and stretches across holes in membranes so thin that they can flex. The grid becomes a thermal landscape with hills and valleys. The metal bars act as heat sinks. The carbon film is thinner and less conductive. The center of a hole is farthest from solid support and is mechanically the most delicate region. When cooling comes, these differences translate into uneven freezing conditions across a single grid square. This is one reason crystalline ice can appear even when the overall procedure is correct: parts of the grid cool differently, and the water in each micro-region experiences its own history.

Then comes the fall into cold. The grid is plunged rapidly into a cryogen, a liquid cold enough to pull heat out of the thin film so quickly that water does not have time to arrange into a crystal lattice. The detail that often surprises newcomers is the choice of cryogen. Liquid nitrogen is colder than liquid ethane, so why not simply plunge into nitrogen? Because temperature is not the whole story; heat transfer is. When a warm object hits liquid nitrogen, the nitrogen at the surface can boil violently and form an insulating vapor layer—an instance of the Leidenfrost effect, the same phenomenon that lets a drop of water skitter across a hot pan on a cushion of steam. A vapor blanket is disastrous when you need rapid cooling, because it slows heat extraction exactly when you want it fastest.

Liquid ethane, although less cold than nitrogen, can cool the specimen more effectively because it wets the grid and does not form such a stubborn insulating layer under typical conditions. Ethane can therefore impose a higher cooling rate on the thin film, making vitrification more reliable. In many setups, ethane is itself cooled by contact with liquid nitrogen, held near its freezing point in a small cup. The apparatus can look like a piece of workshop hardware—clamps, plungers, a metal bath—yet it embodies a deep understanding of heat flow: the goal is not to achieve the lowest temperature on paper, but to pull energy out of a microscopic film before nucleation and growth can take hold.

Even then, perfection is not guaranteed. High-resolution single-particle work has even been reported with samples vitrified in boiling nitrogen under certain conditions, a reminder that the "critical cooling rate" is not a single magic number carved into the properties of water. It depends on what fraction of crystalline ice is tolerable or even detectable, on film thickness, on solutes, and on the subtle ways a specimen and its support exchange heat. Such results do not overturn the practical superiority of ethane for most routine work; they underline a more general point: the preparation is a contest between kinetics and geometry, and the winner can change with details that seem, at first, too small to matter.

The thin film does not merely have to vitrify; it has to do so in a way that produces usable images. Imagine two extremes. In one, the ice is thick and uniform, a smooth pond. Your molecules are safely embedded, but they appear faint, washed out by the amount of water the beam must traverse. In the other, the ice is ultrathin and variable, like a patchy sheen after rain. In some holes there is barely enough thickness to contain the molecules; in others the film has torn or become so thin that the proteins are pushed to the surface. Somewhere between those extremes is a narrow band of success: thin enough for contrast and fast cooling, thick enough to avoid severe interface problems and to accommodate the size of the particles.

This narrow band is why practitioners often speak about "finding good holes." On a single grid, some holes may contain beautifully vitrified ice of the right thickness, while others are unusably thick, broken, or crystalline. It can feel capricious until you remember the grid's thermal landscape. Grid bars cool quickly; regions close to them may vitrify more readily. Centers of large holes may cool more slowly; stresses during freezing can deform the film. A support that looks uniform to the naked eye can be a mosaic of micro-environments under the beam.

There is also a time window that has nothing to do with freezing itself. Between blotting and plunging, the sample sits as a thin film exposed to air. Evaporation can thin it further. The composition can change, especially near edges where evaporation is strongest. Some laboratories control humidity and temperature tightly around the plunge freezer to slow evaporation and improve reproducibility. That need for environmental control is not fussiness. It is recognition that when a film is only tens of nanometres thick in its most useful regions, losing a few nanometres to evaporation is no longer negligible—it can be the difference between a particle floating freely and a particle pinned to an interface.

The lived experience of this preparation step is often a mixture of routine and suspense. A researcher may prepare multiple grids in a session, varying blot times by fractions of a second, adjusting sample concentration, changing support type, or adding small amounts of detergent or other additives to influence how proteins behave at the air–water interface. There is a practical ingenuity here that does not always make it into the celebration of high-resolution maps. The final reconstruction may look like pure computation and physics, but its foundation is often a set of small, hard-won empirical choices made by someone staring at a screen of test images late at night.

A historical anecdote helps place that ingenuity in context. When vitrification first became feasible for hydrated specimens, it did not immediately solve everything. Early grids were difficult to prepare reproducibly, and convincing skeptics required more than a clever idea; it required showing that the "good" state of water could be achieved routinely enough to support a method. Dubochet's early work in the 1980s was influential not only because it introduced vitrified water as a medium, but because it treated the physical state of ice as something you could verify. The ability to distinguish vitreous ice from cubic or hexagonal crystalline forms—by diffraction signatures, by image texture—turned preparation from alchemy into a craft with diagnostics. The field learned to ask, at each step, not "is it frozen?" but "what kind of frozen is it?" That shift in question is one of the quiet revolutions behind the later explosion of structures.

The grid itself, with its perforated film, can be thought of as a device for managing competing demands. It must be conductive enough to reduce charging and to provide a path for heat. It must be mechanically stable enough that a thin film can span holes without tearing during blotting and plunging. It must be compatible with the chemistry of the sample. Even the size and spacing of holes are design choices with consequences. Smaller holes can support thinner films more robustly, reducing the tendency of the ice to bulge or rupture. Larger holes offer larger viewing windows but can encourage greater film sagging and thickness variation. It is easy to forget that "holey carbon" is not one thing but a family of products: many geometries exist, tuned to different particle sizes and behaviors, each embedding assumptions about what the user will want.

A concrete example makes the constraints intuitive. Consider a large ribosomal complex versus a small soluble enzyme. The ribosome is bulky; it needs a thicker region of ice simply to fit without being compressed against interfaces. If the ice is as thin as what might suit a small enzyme, the ribosome may be excluded from the thinnest regions or forced into orientations that reflect surface interactions rather than its free state in solution. For the small enzyme, that thicker ice is a disadvantage: it reduces contrast and makes particles harder to detect and align. The grid does not know what you are studying; it offers a range of thicknesses across many holes, and part of the art is to tune preparation so that a useful fraction of those holes match the biology.

That tuning extends to concentration. Too dilute, and you may find only a few particles per hole, wasting microscope time and making computation less robust. Too concentrated, and particles overlap, clump, or form preferred interactions. Because the water film is thin, concentration in the final ice can be quite different from concentration in the initial droplet. Blotting can preferentially remove bulk liquid while leaving a boundary layer enriched in proteins near surfaces. This can produce unexpectedly high particle densities even when the starting concentration seems modest, another way in which intuition based on test tubes fails when translated to nanometre-thick films.

The plunge itself has a physical drama that is easy to underestimate. The grid accelerates into the cryogen; the film experiences a near-instantaneous temperature collapse. Water's viscosity rises sharply, mobility plummets, and the random arrangement of molecules is trapped into a glassy solid. If the cooling is just a little too slow in a region, crystals can nucleate and grow. Those crystals need not fill the entire hole to cause trouble. A small crystalline patch can generate strong diffraction, and a crystalline front can displace solutes locally as it advances. The difference between success and failure can be local, not global, which is why a grid can contain both excellent and useless areas.

After freezing, the grid is kept cold, transferred into a holder, and eventually into the microscope. By then the decisive act has already happened. The molecules are locked in place, but they are locked into whatever distribution and orientation the preparation step permitted. If particles have all migrated to the air–water interface and adopted similar views, no amount of later computation can conjure the missing orientations. If the ice is too thick everywhere, the images may be too faint. If crystalline ice is widespread, the structured background will swamp the specimen signal. The preparation step is therefore not a prelude; it is the moment when the sample becomes the experiment.

What remains most striking is how the method depends on something that is both exquisitely engineered and almost tactile. A piece of metal mesh, a film only a few nanometres thick, a brief contact with filter paper, and then a plunge into a liquid chosen not for its romance of cold but for its efficiency in stealing heat. The success of the entire enterprise can hinge on a layer of water so thin that it is comparable to the size of the molecules suspended within it, frozen so quickly that water does not have time to become the ice we recognize. In the best regions of a well-prepared grid, the result is a transparent, glassy sheet where proteins sit as if caught mid-drift, separated from one another, hydrated, and held still—ready to be seen.

2.3 The Beam That Will Not Sit Still

A well-made vitrified specimen has a reassuring feel to it: the water is solid, the molecules are trapped, the grid is locked into a cold metal holder. At human scale, everything looks fixed. The microscope column hums, pumps keep the vacuum, and nothing appears to be moving at all.

Then the electrons arrive.

The electron beam is not a flashlight. It is a stream of charged particles that interact strongly with matter. Even at the low doses used for biological specimens, the beam deposits energy, knocks atoms out of place, breaks chemical bonds, and leaves behind a wake of altered material. That part is familiar in outline: radiation damage sets a hard ceiling on how many electrons can be used before the structure is erased. The less intuitive problem is that the specimen can also *move*—not by micrometres, not even by nanometres in an easily predictable direction, but by tiny, jittery shifts that are fatal to high resolution.

The tragedy is that the movements are often smallest in absolute terms when they do the most harm. A blur of a fraction of a nanometre is enough to smear atomic detail. It is like trying to read the fine print on a legal document while the paper trembles in your hand. The letters do not need to leap across the page to become illegible; they only need to waver by an amount comparable to their own stroke width.

Early practitioners could produce samples with vitreous ice and still find that images were disappointing: soft, low-contrast, and stubbornly limited in the detail they would yield. It was natural to blame the microscope, or the complexity of biological particles, or the mathematics of averaging. A quieter culprit was hiding in plain sight: the specimen itself was shifting under illumination. When thousands of images were averaged together, the finest features cancelled out, not because they were absent, but because they were being moved around during exposure.

The problem has several contributors, and it helps to picture them as a set of coupled instabilities rather than a single wobble.

One is mechanical drift: slow motion of the stage or holder. Even in a well-engineered instrument, cooling creates thermal gradients and stresses. Materials contract. Tiny relaxations continue over time. Modern stages are marvels of stability, but they are not metaphysically immobile. Drift is often gradual enough that it can be detected and corrected, and in practice it is no longer the main villain it once was. Still, it sets the scene: the specimen is not sitting on a granite table in a quiet room; it is perched in a cold, pumped column where minute forces can accumulate.

A second contributor is charging. The beam injects electrons into the specimen and causes the ejection of secondary electrons. If those charges cannot flow away efficiently, regions of the specimen can build up electric fields. Electric fields exert forces—on the beam itself, bending its path, and on the specimen, tugging charged regions in subtle ways. Charging can create image distortions that resemble motion blur even when the physical specimen has barely shifted. It can also drive real movement if electrostatic forces push and pull on an ultrathin, weakly supported film.

A third contributor, and the one that gave the field a particularly nasty surprise, is beam-induced motion: the specimen's own response to being irradiated. In single-particle work, many molecules sit in vitreous ice spanning holes in a perforated support. That geometry is wonderfully clean for imaging and wonderfully fragile mechanically. The ice is a thin solid sheet suspended over empty space. The support foil around it can flex. When the beam hits, the combined ice-and-foil system can deform, sometimes in a way that resembles the skin of a drum responding to a tap. A detailed mechanistic picture has emerged in which the foil and the ice layer undergo a "drum-like" deformation mode during exposure, moving particles in a coordinated but spatially varying pattern. A particle near the centre of a hole may shift differently from one near the edge; two particles in the same micrograph can drift in slightly different directions.

This was not merely speculation. Work on vitrified specimens on holey carbon highlighted that beam-induced motion could be a major source of blur in single-particle data, and that certain experimental levers made it better. Smaller holes tended to reduce the motion. Lower dose rates could help as well. Those observations point away from the microscope stage as the sole origin and toward something local and mechanical: if the hole is smaller, the suspended "drumhead" is stiffer; if the beam is gentler per unit time, the stresses may relax differently.

The emphasis on *dose rate* introduces one of cryo-EM's most uncomfortable balancing acts. Radiation damage cares primarily about total dose: the number of electrons per unit area that have passed through. If you exceed the sample's tolerance, fine structure disappears into chemical wreckage. But motion can care about how quickly that dose is delivered. A high dose rate can trigger rapid charging and faster deformation. A low dose rate can reduce some of those instabilities, but it also forces longer exposures, giving drift more time to accumulate and making the image more vulnerable to any slow movement. There is no setting that turns the specimen into a perfect photographic subject. Every choice trades one source of blur against another.

Low-dose imaging, necessary to stay within the damage limit, makes everything harder in an everyday way: it makes the images noisy. With light photography, you can shorten exposure time to freeze motion and simply add more light. Here, "more light" is more electrons, and more electrons destroy what you are trying to record. So exposures are kept low, and the raw images are faint. This faintness used to force people into a compromise that now feels archaic: taking one long exposure and hoping the specimen stayed still enough to retain high-resolution information.

A change in detector technology—direct electron detectors—helped break that compromise by making it practical to record data as movies: a rapid series of frames, each receiving only a fraction of the total dose. The idea is called dose fractionation. Instead of one blurred photograph, you take many dim ones. If the specimen has moved slightly between frames, you can align the frames computationally and add them together in register, recovering sharpness that would otherwise be lost.

This approach redefined what it meant to "take an image." The exposure became a time-resolved event, and the data contained a record of the specimen's restlessness. That record brought its own revelations. Motion is often worst at the beginning of exposure. The earliest frames contain some of the least radiation-damaged information, the part you would most like to preserve, but they can also carry the largest sudden shifts as the specimen responds to the initial blast of electrons. Later frames may be steadier, but their high-resolution signal is progressively erased by damage. The best information is therefore trapped between two forms of loss: it is born into motion and dies into damage.

Once movies became common, motion correction turned into a central computational task. Early correction treated the entire micrograph as a single rigid object, shifting all frames by the same amount to compensate for global drift. That helps when the whole specimen slides uniformly, but it fails when different regions move differently, as they can when ice and foil deform locally. A major advance was the development of algorithms that correct not only for whole-frame drift but also for spatially varying motion across the image.

MotionCor2, introduced by Shaoqing Zheng, Eugene Palovcak and colleagues in 2017, became widely used because it addressed this "anisotropic" motion: it could model and correct local shifts that differ across the micrograph. The conceptual shift is important even for readers who never touch the software. It says that motion is not just an annoyance that smears everything equally; it is a patterned physical phenomenon tied to the mechanics of thin films and supports. Once you treat it that way, you can begin to recover signal that was genuinely present but displaced.

The fact that such recovery is possible is both hopeful and unsettling. Hopeful, because it means earlier datasets contained more information than was realized; unsettling, because it implies the specimen was always moving and the field simply lacked the tools to admit it. There is a common rhythm in experimental science: first you blame your apparatus, then your sample, then your mathematics, and only later do you learn which of those was actually limiting you.

Charging adds another layer of complexity because it can masquerade as motion and also amplify it. Conductivity becomes valuable not only for its electrical quietness but for its mechanical implications as well. If a specimen region cannot shed charge, it can develop local fields that pull on the thin ice film or deflect the beam. That is one reason why researchers have explored adding conductive layers such as graphene. Graphene—a single-atom-thick sheet of carbon—can provide a pathway for charge to dissipate while adding very little background scattering. In favourable cases it mitigates charging with minimal additional noise, which is a remarkable bargain at the scale where every extra atom along the beam path counts.

At this point it may be tempting to imagine an engineer's simple fix: make everything stiffer, more conductive, more stable, and the problem goes away. The reality is trickier because the specimen is a composite object. It includes metal grid bars, a perforated support film, an ice layer spanning holes, and embedded macromolecules. Each has different thermal expansion, conductivity, and mechanical stiffness. Cooling introduces stresses; the beam introduces further stress and charge. The specimen is not a passive slide but a small machine being driven out of equilibrium.

The human experience of this problem is often strangely intimate. A microscopist can watch it happen. In a modern instrument, the camera displays the drift and defocus changes in real time. In movie mode, one can inspect frames and see that particles have shifted during exposure. When someone says "this grid is moving," they do not mean the entire holder is sliding across the lab bench; they mean the ice in a hole is flexing by ångströms in response to an invisible force, and that those ångströms will decide whether a side chain is resolved or blurred into anonymity.

There is also a psychological shift that comes with accepting beam-induced motion as central. It reframes resolution as something that is continually lost rather than something simply provided by the microscope. The instrument may be capable of forming an image with extremely fine detail, but the specimen and support can erase that detail before it is recorded. Resolution becomes a negotiated outcome between the desire to use enough electrons to see a faint object and the need to keep that object both chemically intact and mechanically quiet.

Concrete consequences show up quickly in data collection strategy. If motion is largest at the beginning, some processing pipelines down-weight early frames or treat them differently, attempting to preserve their high-resolution content while accounting for their instability. If dose rate affects motion, one might prefer longer exposures at lower instantaneous intensity, even though it demands patience and more careful drift management. If smaller holes reduce deformation, one might choose a support with a different hole size or pattern, accepting other trade-offs in particle distribution or ice quality. Choices that once seemed like mundane consumables—hole geometry, foil material, support treatment—become part of an effort to keep the specimen from flinching.

Perhaps the most poignant aspect of beam-induced motion is that it is an unwanted form of communication between the instrument and the specimen. The beam is supposed to be a probe, a way of asking the sample what it looks like. Instead, it becomes a hand that nudges the object while trying to sketch it. The nudges are microscopic, but the sketch is so demanding that even microscopic nudges leave a visible trace.

Yet the story is not one of defeat. The reason motion became a central enemy is that it was both ubiquitous and, to a surprising extent, tractable. Once movies exposed the problem, and once algorithms like MotionCor2 learned to align not just frames but local regions, sharp information could be recovered from data that would previously have been dismissed as hopelessly blurred. The improvement was not only numerical, in reported resolutions, but visceral: features that had been hinted at in fuzzy densities began to appear as clear strands and ridges, as if someone had finally tightened the focus on a restless subject.

A frozen specimen, it turns out, is not truly still. It is merely slow enough that with the right combination of gentle illumination, fast recording, and careful correction, its small motions can be tracked and undone—leaving behind an image that is less a snapshot than a carefully stabilized portrait.

2.4 Quiet Metals, Better Pictures

A microscope is often described as if it were a single object: a column, a lens system, a camera. In practice it is an arrangement of bargains. Every gain in sensitivity brings you closer to the specimen's fragility; every attempt to stabilize one part of the system reveals another part that flexes. After the discovery that the beam can make a frozen specimen shift, it became hard to unsee the sample holder as a component of the imaging system rather than a neutral platform. The grid and its supporting film were not just a stage; they were part of the plot.

This is a slightly humbling realization. The materials involved can look like the least glamorous components in the room: tiny discs of metal mesh, thin foils with a pattern of holes, films that can be damaged by a careless touch. Yet these "consumables" sit at the junction between the macroscopic world of the instrument and the microscopic world of the specimen. They conduct heat away during freezing, conduct charge away during irradiation, and resist (or fail to resist) the mechanical strains that the beam provokes. When they misbehave, they translate invisible forces into blur.

For years, carbon was the default material for support films. Amorphous carbon has virtues that are easy to appreciate: it is relatively easy to make into thin films; it is compatible with many surface treatments; and it is reasonably well behaved in the vacuum of the microscope. It can also be a source of problems that are easy to overlook. Carbon films can charge under the beam, building electric fields that distort images and may tug on the ice. They can also flex. A suspended region spanning a hole is not a rigid plate; it is a membrane under tension, and the beam's energy can turn that membrane into a barely audible drumhead.

The phrase "beam-induced motion" can make the problem feel abstract, as if it were an unavoidable curse attached to charged particles. Materials engineering showed that much of the motion is not ordained; it is mediated. Change the support and you can change the specimen's response. The key is to make the support *quiet*: mechanically stiff, electrically conductive, and thermally well behaved at cryogenic temperatures.

Gold turned out to be an unusually good candidate for this quietness. Gold is an excellent conductor, so charges have a pathway to drain away rather than piling up as localized electric fields. It is also dense and mechanically stable in thin foils. And it is chemically inert enough to be a practical material in a laboratory environment. Gold had been present around microscopes for decades in the form of coatings and contacts, but its role as a support for high-resolution cryo-EM was not initially obvious. A support has to do more than be stable; it has to be manufacturable into a perforated film that remains flat, survives handling, and provides good ice.

A particularly influential demonstration came from Christopher Russo and Lori Passmore in 2014, who reported ultrastable gold substrates for cryo-EM and showed that they could reduce beam-induced motion compared with more traditional carbon supports. The effect was not a small cosmetic improvement. Reduced movement means that high-resolution information survives longer into the exposure, which means that fewer particles are required for a given resolution target and that reconstructions become more routine rather than heroic. The beauty of the result was its directness: change a material, and the images sharpen. No new physics of scattering was required, only an appreciation that the specimen-holder assembly is an active participant in the experiment.

The everyday adoption of gold-foil grids followed. Commercial supports such as UltrAuFoil-type grids are used in practice because their stability reduces beam-induced motion and improves data quality. There is a quiet sociological aspect to this kind of change. When a technique becomes widely used, it is often because it reduces the number of "bad days" in the microscope room. A better support does not merely allow a record-breaking dataset; it raises the floor, producing usable micrographs more often, with fewer hours spent wondering whether the problem is your sample, your alignment, your microscope, or your luck.

To understand why the support material can have such an outsized effect, it helps to return to the composite nature of the specimen. A typical single-particle preparation places molecules in a glassy water layer spanning holes in a foil. Under the beam, the foil and ice can deform together, and that deformation can shift the embedded particles by ångströms. Such movements are small in absolute terms and enormous in their consequences. Work describing "drum-like" deformation modes argues that when foil/ice movement is reduced, specimen movement can be pushed toward sub-Å levels. That threshold is psychologically important: below an ångström, you are no longer averaging away most of the atomic detail simply by exposure-induced shifting. The microscope's resolving power and the computational tools for alignment can finally be used closer to their full potential.

Gold helps not only by being stiff and conductive but also by changing how stresses are distributed. A more rigid support can reduce the tendency of suspended regions to bow and relax under irradiation. Conductivity lowers charging, which reduces electrostatic forces that can either deflect the beam or physically tug on the specimen. Thermal properties also play a role, because the specimen is cryogenic and any small amount of beam-induced heating has to be dissipated through the grid and support. A material that provides good thermal contact can reduce localized temperature excursions and the associated mechanical relaxation, though heating in typical low-dose imaging is not the main driver compared with mechanical and electrostatic effects.

Once gold entered the picture, it became natural to ask a further question: can you make the support both quieter and less intrusive? Every support adds background. Dense materials scatter strongly, and although the beam is focused on the holes, no experiment is perfectly isolated from its surroundings. The support is also a surface, and surfaces influence how proteins distribute and orient. The ideal would be a support that is mechanically stable and electrically conductive but adds almost no extra scattering and presents a gentle, well-defined surface chemistry.

This is where graphene entered the conversation. Graphene is a single layer of carbon atoms in a hexagonal lattice: atomically thin, mechanically strong, and electrically conductive. In principle, a graphene sheet is close to the ideal minimal support: it can conduct charge away and provide a continuous surface with very little added material in the beam path. In practice, pristine graphene is chemically hydrophobic and can be difficult to work with; proteins may not spread evenly on it without modification. Yet its combination of conductivity and thinness made it attractive as a tool for reducing charging and stabilizing specimens.

Graphene-on-gold is an especially instructive hybrid because it links two different virtues. Gold provides a mechanically stable, highly conductive foundation. Graphene adds an ultrathin conductive layer that can help with charging while contributing minimal additional background. The pairing is also a reminder that "support" is not one thing. It is a stack: a mesh grid, a foil, sometimes an additional layer, and then the ice. Improvements can be made at multiple levels, and the best performance may come from combinations rather than single-material purity.

There is a deeper lesson here about how experimental science evolves. When people tell the story of a "resolution revolution," they often focus on dramatic intellectual leaps: new detectors, new reconstruction algorithms, new ways of classifying heterogeneous particles. Those breakthroughs were real, but the revolution also depended on making the specimen behave like a cooperative physical object. Quiet supports are part of an engineering feedback loop. Better stability produces better images. Better images make motion more measurable. Once motion can be measured, it can be corrected computationally and, crucially, minimized through further design. The field learns what to fix by first seeing what is broken.

A concrete anecdote from the microscope room captures how these improvements feel. A researcher may spend weeks optimizing a protein sample—purifying it, checking its activity, adjusting buffer conditions—only to discover that the images are plagued by motion blur or charging artifacts. Switching to a more stable support can change the experience overnight. Features in the raw micrographs become easier to detect. The contrast improves not because the protein has changed, but because the background has become better behaved and the particles remain more stationary during exposure. Data collection becomes less of a gamble. Instead of harvesting a few "good" micrographs from a sea of unusable ones, the researcher begins to collect hundreds or thousands of consistently sharp images. That change shifts not only the quality of a single project but the kinds of projects a laboratory dares to attempt.

Quiet supports also change the emotional texture of the work. When high-resolution results are rare, each promising dataset carries a sense of fragility: any small misstep could waste days of precious microscope time. More stable supports make success less episodic. They do not eliminate the need for expertise—sample preparation remains an art—but they reduce the frequency with which physics sabotages even well-prepared specimens.

Not all the surprises point in the same direction. A particularly intriguing nuance is the suggestion that the ice itself, not only the metal, can be part of the "quiet" design space. Controlled devitrification has been reported to reduce early-frame motion in some contexts, implying that the stress or phase state of the frozen water can influence how the specimen responds at the start of exposure. The idea is counterintuitive: the glassy state is prized because it avoids crystalline diffraction and preserves structure, yet a carefully managed departure from the ideal glass may relieve stresses that otherwise drive motion. The point is not that crystalline ice is suddenly desirable—it remains a destructive presence—but that the mechanical state of the frozen film can matter as much as its diffraction signature. Water's physics continues to participate long after freezing is complete.

That nuance also hints at why motion remains a bottleneck in high-resolution work, even as detectors and algorithms improve. Reviews of the field still frame motion as a limiting factor, which suggests that no single fix will abolish it. Quiet metals reduce one set of problems, but motion is a compound phenomenon tied to charging, mechanical relaxation, radiation damage, and the geometry of suspended films. Improvements come from stacking advantages: a stable support, good electrical conductivity, appropriate hole size, careful control of ice thickness and stress, and computational correction that can track what remains.

Gold supports have, in a sense, altered the baseline expectation. They have made it easier to demand that the specimen be still, and easier to notice when it is not. That is what successful engineering does: it turns an unavoidable-seeming source of failure into a solvable problem, and then into a new standard. A gold grid does not look like a breakthrough in the way a new microscope might. It is small, quiet, and easily overlooked. Yet its contribution is written into the sharpness of modern reconstructions, in the recovery of fine features that would otherwise be smeared away by a film that flexes or a surface that charges.

The molecules embedded in frozen water cannot speak, but they can be made audible. Sometimes the most effective way is not to shout with more electrons, but to build a stage that does not tremble when the spotlight turns on.

Chapter 3

Inside the Cathedral of Electrons

At first glance, a cryo-electron microscope looks less like a camera than a shrine built to appease unruly physics: vacuum, cold, silence, and exquisite alignment. Step by step, the reader discovers how electrons are steered, how fleeting signals are rescued before radiation destroys the specimen, and how an instrument once reserved for virtuosos became a high-throughput engine for molecular sight.

3.1 A Machine Built Against the World

The first shock of meeting a modern cryo-transmission electron microscope in person is that it does not resemble a camera so much as a building service. It rises like a tall, white column anchored to the floor, surrounded by consoles, pumps, cable looms, and warning lights. The room around it is often as carefully engineered as the instrument itself: thick concrete, controlled temperature, filtered air, sometimes an entire floor poured as a single slab to keep the ground from shivering. The paradox is the point. This is a machine built on a heroic scale to protect something vanishingly small: a few faint traces left by electrons after they have skimmed through a film of frozen water holding molecules in place.

At the heart of the instrument is a simple ambition: send a beam of electrons through a specimen and record the pattern of scattering. Everything else is defensive architecture. Ordinary reality—air currents, warmth from a human hand, the tremor of footsteps, a smear of hydrocarbon residue, a whisper of frost—can overwhelm the information in a single exposure. If a cryo-TEM occasionally feels like a cathedral, it is not because it is mystical. It is because it demands a kind of hush.

The electron beam begins where the column begins: at the source, often called the *gun*. In a light microscope, photons are plentiful, forgiving, and easy to guide through glass lenses. Electrons are none of those things. They do not travel through air for any useful distance without colliding and scattering. They must be accelerated to high energy and coaxed through a near-empty column. This requirement shapes the instrument from top to bottom: a tall vacuum tube with electromagnetic lenses spaced along it like stations in a vertical railway. The specimen sits partway down, and the detector waits near the bottom, where the beam finally arrives carrying whatever the specimen has done to it.

Even the word *lens* is a little misleading here. In a cryo-TEM, focusing and steering are done by coils of wire and carefully shaped iron pole pieces that generate magnetic fields. These fields act on electrons the way glass acts on light: they bend trajectories, bring rays to a focus, and form an image. The catch is that magnetic lenses are exquisitely sensitive to small instabilities. A slight drift in current, a tiny temperature change that expands a component, a minute mechanical relaxation in the stage, or a weak external magnetic field can shift the beam. When the details you care about are measured in angstroms, a shift that would be unnoticeable in most machines becomes the difference between seeing a crisp edge and seeing a blur.

Start with air, the most basic adversary. The column is pumped down to high vacuum so electrons can fly unimpeded. Yet the instrument must also accept a specimen that began its life in water, was plunged into a cryogen to lock it in place, and is then transferred into the vacuum without thawing and without picking up ice crystals from the atmosphere. That combination—wet origins, deep cold, and a vacuum environment—creates a unique kind of fragility. A drop of ambient humidity is no big deal on a laboratory bench. Inside a microscope column, on a sample held far below freezing, it becomes frost: an invisible snowfall that can bury particles, thicken the ice, and add unwanted scattering long before any electron reaches the detector.

The old story of cryogenic imaging is full of such practical battles. Jacques Dubochet and colleagues, in work that crystallized into a landmark 1988 paper, demonstrated that biological material could be embedded in water frozen so quickly it solidified into a glassy state rather than crystalline ice. That achievement did not merely preserve structure; it created a way to carry hydrated molecules into a vacuum. But it also set a trap for every future practitioner: glassy ice is stable only if it stays cold and clean. Warm it even a little and it begins to relax toward crystalline order. Expose it to humid air and it becomes a magnet for frost. The very trick that makes the specimen possible also makes it demanding.

The specimen's journey into the microscope is therefore choreographed. It travels on a metal cartridge or *holder* designed to maintain cryogenic temperature and to dock into the stage without letting room air reach the grid. Some instruments use automated loaders: a robotic arm moving grids between a storage dewar and the column through a series of vacuum locks. The goal is always the same. Every second spent at the wrong pressure or the wrong temperature increases the odds that water molecules will redeposit as crystalline contamination. People who work with these instruments develop a special kind of attentiveness: a sense for how long a transfer can safely take, how quickly a door should be opened and closed, how much patience to have with a pump-down cycle. In cryo work, time is not money so much as cleanliness.

Inside the column, the vacuum has its own personality. A vacuum is not an absence so much as a controlled emptiness, always threatened by slow leaks, outgassing, and the chemistry of whatever has been introduced. Hydrocarbons are a particular menace. Minute traces from pump oils, lubricants, fingerprints, or plasticizers can migrate and condense under the beam, especially on cold surfaces, building up an amorphous film that adds noise and changes contrast. Microscopists speak about contamination the way cooks speak about smoke: you may not see it at first, but you can taste it in the final result. There are even characteristic visual artefacts that become part of the shared folklore of facilities, such as patchy, mottled ice patterns sometimes associated with subtle column problems. These are not just aesthetic blemishes; they are reminders that the instrument is an ecosystem.

Heat is the next ordinary force that must be defeated. The specimen is held extremely cold to preserve its structure and to slow the chemical rearrangements that electron exposure can trigger. But cold creates gradients, and gradients create motion. Any object held at cryogenic temperature inside a warmer environment becomes a point where gases condense. It also becomes a mechanical system under stress. Metals shrink as they cool; different parts shrink by different amounts; contacts tighten or loosen; lubricants behave differently. A stage that positions a grid with nanometre precision is asked to do so while one part of it is cold enough to freeze air and another is near room temperature. The engineering is not simply about reaching a low temperature, but about reaching it without introducing a slow creep that continues over minutes and hours.

That creeping motion has a name in the microscope world: *drift*. Drift is not dramatic. It is the gentle, persistent sliding of the specimen relative to the beam during an exposure. Drift can come from mechanical relaxation after the holder is inserted, from thermal equilibration, from charging effects under the beam, or from tiny instabilities in the stage mechanism. At low resolution, drift may barely be noticed. At high resolution it is lethal, because it smears fine detail into a haze. One of the most sobering facts about high-end imaging is how little movement is required to ruin it. If the structural features you want to resolve are separated by a few angstroms, then motion of comparable size during the exposure wipes them away. Magnification does not save you. It only magnifies the blur.

Vibration is drift's louder cousin. A microscope column is tall, and tall structures have resonances. The floor has resonances too, and so does the building. A distant elevator, a passing truck, a subway line a block away, the thump of a door, the footfall of a person shifting their weight: all can couple into the system. Facilities learn this the hard way. There are stories, passed along with equal parts amusement and pain, of instruments installed in rooms that seemed perfect until the first dataset revealed a persistent, rhythmic blur. Someone eventually correlated it with the building's air-handling cycle, or with a pump switching modes, or with delivery vehicles that arrived at the same hour each morning. The microscope had not changed. The world had.

To counter this, the instrument is often mounted on vibration-isolation systems. Some are active, using sensors and actuators to cancel motion; others are passive, using mass and damping to shift resonant frequencies away from the most harmful range. The room is kept at stable temperature because air expands and contracts, and because electronics drift when they warm. Even the choice of where to put a microscope in a building can become a scientific decision. A basement is stable but may sit on a rumbling foundation. An upper floor is quiet but can sway. The most expensive microscopes sometimes end up in rooms that resemble recording studios more than laboratories, designed for silence not of sound but of motion.

Magnetic fields are another mundane intrusion. A cryo-TEM relies on magnetic lenses with fields far stronger than Earth's, but it is precisely because the instrument is magnetic that small external fields can be problematic. Nearby elevators, power cables, or other heavy equipment can generate fluctuating fields that slightly deflect the beam or shift the lens conditions. Some rooms use magnetic shielding; others rely on distance and careful routing of electrical systems. This is one reason facilities like to control what can be wheeled into a microscope room. A seemingly innocent cart with a large motor, parked too close for convenience, can alter an image in ways that are maddeningly intermittent.

All this effort is spent to preserve a signal that is intrinsically stingy. Biological material in frozen water does not strongly absorb electrons; it mostly nudges their phase and direction. The instrument must therefore be as efficient as possible at turning tiny scattering differences into a measurable image. That starts with the electron optics. The *condenser* lenses above the specimen shape the illumination: how broad the beam is, how parallel it is, how much current it carries. The *objective* lens, the most important lens in the instrument, sits around the specimen region and forms the first image. Downstream lenses magnify that image and project it onto the detector. Each stage offers opportunities for aberrations, slight imperfections that distort the image. In light microscopes, aberrations are familiar—spherical aberration, chromatic aberration, astigmatism. Electron lenses suffer analogues, and because they are electromagnetic, they also inherit sensitivities to currents, temperatures, and mechanical tolerances.

Astigmatism provides an intuitive example. Ideally, a lens focuses equally in all directions. If it focuses more strongly along one axis than another, a point becomes a line, and fine detail becomes directionally smeared. Correcting astigmatism is partly a matter of tuning the instrument with additional coils called stigmators, but tuning is only meaningful if the underlying conditions are stable. The microscopist becomes both photographer and custodian, adjusting alignments while watching for the signs that something deeper is shifting: contamination creeping in, a vacuum level changing, a stage that has not yet settled.

The specimen stage itself is a small mechanical miracle. It must move a grid across micrometres to find suitable areas while also holding it steady to within angstroms during exposure. It must tilt for certain kinds of imaging while maintaining eucentricity, the geometric condition that keeps the area of interest near the same point as the stage tilts. It must do all this while connected to a cryogenic holder or autoloader mechanism that brings cold into the column without bringing air. Every coupling—mechanical, thermal, electrical—is a path for noise to enter. The stage is therefore engineered as a compromise: stiff enough to resist vibration, thermally stable enough to limit drift, accessible enough to be practical, and precise enough to be worth the effort.

The detector at the bottom of the column completes the chain. It receives electrons that have survived the journey, carrying the delicate imprint of the specimen's scattering. Yet even before detection, the electrons can be betrayed. If the column is not clean, electrons scatter off contamination and add background. If the beam is not stable, the illumination varies across the exposure. If the specimen accumulates charge, it can act like a tiny lens of its own, bending trajectories unpredictably. If the ice layer is too thick, multiple scattering events wash out high-resolution information. The instrument is therefore judged not merely by whether it *can* produce a sharp image, but by how often it can do so under real working conditions, with real samples that do not cooperate.

One of the most vivid ways to appreciate the cryo-TEM's hostility to the everyday world is to consider how it treats water. In ordinary life, water is the universal solvent, the friendly background of biology. Here it is both the medium of preservation and a potential saboteur. A thin film of frozen water is essential for holding molecules in a near-native arrangement, but the same water, if it crystallizes, becomes a diffraction grating that stamps its own strong pattern onto the image. Water vapor in the air, if it condenses, becomes frost that hides the particles. Even water adsorbed onto surfaces inside the column can migrate and redeposit. The microscope is constantly negotiating with water: allowing enough to preserve biology, forbidding enough to preserve the image.

The result is an instrument that feels less like a passive observer than a controlled environment with a beamline attached. It is designed to make electrons behave predictably, to make the specimen behave inertly, and to keep the rest of the universe from interfering long enough to take a picture. People who master it often sound less like they have learned a single technique and more like they have learned a way of living with a demanding machine. They learn to listen for the pitch of a pump, to recognize the subtle change in image texture that signals contamination, to wait out the slow settling of drift, to treat a cold grid as if it were both fragile and precious. The instrument rewards that care not with spectacle but with a kind of quiet fidelity: the ability to hold still, for a moment, the patterns that molecules make.

3.2 When the Camera Learned to Count

A cryo-TEM can be tuned into an almost stubborn stillness: vacuum steady, stage settled, lenses aligned, the specimen held cold enough that water behaves like a solid glass. Yet for decades, the final step—the moment when scattered electrons were turned into a picture—was oddly wasteful. It was as if someone had built an observatory on a mountaintop and then insisted on peering through a fogged window.

The problem was not ignorance. It was that the detector, the microscope's "camera," had to do something fundamentally difficult: register an extremely small number of electrons without adding much noise of its own. In biological imaging the dose budget is tight, so every electron counts in the plain sense of the word. If half of them are lost in the detector, or if the detector adds a hiss of randomness on top of the signal, then no amount of careful sample preparation can rescue the information. A blurry photograph of a rare bird is still a blurry photograph, even if the bird is perfectly still and the lens is perfect.

For much of the twentieth century, the most faithful recording medium was simply film. Electrons struck a phosphor screen, which converted their impacts into light, and the light exposed photographic emulsion. Film had real virtues: large area, fine grain, and a kind of straightforward honesty. It was also slow and inconvenient. Developing, scanning, and managing film created a bottleneck, and the workflow encouraged a style of microscopy that was artisanal and scarce: each exposure felt precious, each adjustment a kind of wager.

When digital detectors arrived, they did not immediately solve the most important problem. Early digital approaches typically used a conversion layer: electrons hit a scintillator that emitted light, and the light was then sensed by a CCD (charge-coupled device). This made images easier to capture and manage, but the very act of conversion blurred the signal. Imagine trying to learn a melody by listening through a wall. The tune is there, but the wall smears sharp notes into a muffled hum. Scintillators spread the energy of each incoming electron into a little splash of light, and the optics that carry that light to the sensor add their own blur. The result is that fine detail is attenuated, even if the microscope itself could, in principle, deliver it.

Microscopists have a way of describing this with a term borrowed from the language of measurement: *detective quantum efficiency*, usually shortened to DQE. The name is forbidding, but the idea is friendly. DQE asks: if a certain amount of information arrives at the detector in the form of electrons, how much of that information survives the act of detection? A perfect detector would preserve the signal-to-noise ratio. A flawed detector would throw away some information and add its own noise. Under the strict dose limits of biological imaging, that loss is not a small inconvenience; it is a hard ceiling.

The turning point came when detectors were built to sense electrons *directly*, without translating them first into light. These are called direct electron detectors. They are made from semiconductor sensors engineered so that an incoming high-energy electron creates charge that can be collected and measured in the pixel it strikes (or near it). Suddenly, the "fogged window" was replaced with a clean one. The improvement in DQE meant that images did not just become prettier. They became more truthful at high spatial frequencies, the fine patterns corresponding to small structural details.

But a second change mattered just as much as the first. Direct detectors were not only more efficient; they were fast. They could record images as a rapid series of frames. That seems like a feature borrowed from consumer electronics—a high frame-rate camera shooting a video—but in cryo work it transformed what it meant to take a single exposure.

Older cameras encouraged a single long integration: open the beam for a chosen time, collect whatever happens, close the beam. The exposure becomes a kind of averaging over the entire event. If the specimen shifts slightly, or if it bends under the beam, or if the stage drifts, the final image contains that motion as blur. There is no way to disentangle "what the specimen was" from "how it moved" because the detector has already committed the whole interval to one picture.

A fast detector changes the logic. Instead of one long exposure, you record a *dose-fractionated* movie: the same total number of electrons, but divided across many short frames. The idea was demonstrated forcefully in the early 2010s, with work by Xueming Li and colleagues (2013) showing how electron counting and motion correction together enabled near-atomic-resolution single-particle reconstructions. Dose fractionation does not create more information out of nothing; it simply refuses to throw information away prematurely. Motion, once baked into a blur, becomes something that can be estimated and corrected frame by frame.

Why does the specimen move at all? Part of the answer is mundane: even a well-engineered stage drifts a little, and a cold holder relaxes after insertion. But there is a deeper, stranger source of motion: the beam itself. When electrons strike the sample, they deposit energy. Even at cryogenic temperature, that energy can cause tiny mechanical stresses, charging, and subtle rearrangements in the supporting ice and the specimen. The motion is often largest at the beginning of the exposure, when the sample is "settling" into the new state imposed by the beam. To a human observer, nothing seems to move. To a molecule-sized target, the world is quivering.

Once you have a movie, motion correction becomes possible in an intuitive way. If successive frames show the same object shifted slightly, you can align the frames to one another and sum them, producing an image with less blur. Early approaches treated motion as if the entire field shifted rigidly, like a slide being nudged. That helps, but it is not always realistic. The specimen can flex. Different regions can move in different directions, especially across a hole where the supporting film is absent and the ice behaves like a tiny suspended membrane. This is why motion correction evolved toward local, patch-based strategies that allow different parts of the image to be aligned separately. A widely used implementation, MotionCor2 (Zheng and colleagues, 2017), embodies this practical insight: the sample's motion is not always a single drift vector; it can be anisotropic and locally varying.

A second computational trick becomes natural once you think in frames: not every frame is equally valuable. As exposure proceeds, the specimen accumulates changes from the beam. The earliest frames may be sharp but affected by rapid initial motion; later frames may be more stable but carry less high-resolution detail because the finest features have been progressively degraded. With a movie you can *weight* frames differently, leaning more heavily on the parts of the exposure that preserve fine information and down-weighting those that do not. This is not a moral judgment about damage; it is a form of bookkeeping, an attempt to spend the dose budget wisely.

If dose fractionation was the conceptual shift, electron *counting* was the change in how the detector treated each electron. An integrating detector measures the total charge accumulated in a pixel over an interval. That is simple, but it blurs an important distinction: is a given amount of signal in a pixel produced by one electron depositing a lot of energy, or by two electrons, or by one electron plus noise? Counting detectors take advantage of fast frame rates to treat individual electron impacts as discrete events. Each event is detected, localized, and tallied. This suppresses a major source of noise associated with integrating readout and makes the statistics of the image closer to the underlying physics: a sparse rain of electrons.

Counting brings its own constraint, one that is almost comic in its simplicity. If two electrons arrive so close together in time and space that the detector cannot separate their events, they may be registered as one. This is called *coincidence loss*. Anyone who has tried to count people entering a concert hall knows the feeling: if the crowd comes through the door in a tight cluster, you miss some. In electron counting, coincidence loss means that the dose rate must be kept low enough, or the frame rate high enough, that events are unlikely to pile up in the same pixel between readouts. The result is a trade-off that affects the rhythm of data collection. You want enough electrons to form an image, but not so many per frame that you break the very counting regime that improves the image.

The benefits of counting and movies are easiest to appreciate with a concrete mental experiment. Picture a small protein complex embedded in glassy ice: faint, low contrast, and surrounded by a background that is only slightly different from the particle itself. Under an older camera, you might take a single exposure long enough to get some contrast. During that time, the particle shifts a little and the finest details smear out. You can still align many such particles computationally and average them, but the highest-resolution information has been softened in every image before processing even begins. Averaging cannot reconstruct details that were never recorded cleanly.

With a direct detector, you can take the same total exposure as a rapid sequence. Now the particle's small motions are visible as frame-to-frame shifts. You can align the frames to undo much of that blur. Because counting improves the effective signal-to-noise ratio, each frame is less compromised by detector noise. Then, when many particles are combined, the reconstruction is built from cleaner high-frequency information. The revolution was not a single trick; it was a coupled system: a detector that could count, a readout fast enough to make movies, and software that could use those movies to correct motion and weight dose.

It is tempting to narrate this as a triumph of computation over physics, but that misses the emotional texture of what changed in laboratories. The daily experience of collecting data became less like taking a few precious photographs and more like recording streams. Researchers stopped arguing only about which micrograph was "good" and began arguing about how to manage terabytes of movies, how to automate preprocessing, how to assess motion and image quality on the fly. The microscope room gained an invisible companion: a server rack down the hall, a storage array, a pipeline of software that would begin aligning frames almost as soon as they were saved. The camera did not merely capture better images; it altered the tempo of the entire enterprise.

There is a historical poignancy here. Film, for all its inconvenience, had forced a kind of slowness that matched the difficulty of the specimens. People would spend long hours perfecting alignment and carefully choosing when to expose. Early digital detectors sped up the workflow but did not solve the fundamental scarcity of information at low dose. Direct detectors changed the emotional balance. Suddenly, many targets that had been “almost” visible became tractable. Not because the molecules became more cooperative, but because the instrument stopped squandering electrons.

The phrase “the camera learned to count” captures a genuine change in attitude. Counting is what you do when events are precious. You count rare birds, votes, heartbeats, and coins. A detector that counts is implicitly admitting that each electron is a discrete messenger. That mindset aligns with the dose budget: you have only so many messengers before the specimen is changed beyond usefulness, so you had better listen carefully to each one.

Even with counting and motion correction, the images remain subtle. The raw frames in a movie often look unimpressive to the untrained eye: grey fields with barely visible specks. The transformation happens when hundreds of frames are aligned, summed, and weighted, and when tens or hundreds of thousands of particle images are combined into a three-dimensional reconstruction. Yet the first transformation is already profound: a single exposure becomes not a single frozen moment, but a record of a fragile object reacting in real time to being observed.

A final irony is that the advance brought a kind of humility. It revealed that the specimen had been moving all along. The older cameras simply could not show it. When direct detectors arrived, microscopists began to see beam-induced motion as a dominant blur source, not an occasional nuisance. Once seen, it could not be unseen. That recognition reshaped how people thought about grids, supports, ice thickness, stage stability, and even the etiquette of the microscope room. A door closing at the wrong moment had always been a problem; now it could be measured in angstroms.

In the end, the detector's job is brutally simple: take a sparse, jittery pattern of electron arrivals and turn it into data without lying. Direct electron detectors made that honesty attainable. They did not abolish the dose budget or the sample's fragility. They did something subtler and more enabling: they allowed low-dose imaging to retain fine detail by recording events quickly, efficiently, and with a kind of numerical respect for each electron's existence. When the counting works and the motion is corrected, the resulting images have a new quality—not just sharper edges, but a sense that the molecule is finally being heard over the instrument's own noise.

3.3 Making Shadows Speak

Even with the most stable column and a camera fast enough to count electrons, a raw micrograph can feel like an anticlimax. The specimen is there, but it does not announce itself with bold outlines. Particles hover as faint smudges against a background of frozen water that is only slightly different from the thing you want to see. People encountering their first dataset sometimes assume something has gone wrong: the grid must be empty, the focus must be off, the microscope must be misaligned. Often none of those are true. The image is simply obeying the physics of how biological matter interacts with electrons.

The key twist is that many biological specimens in frozen water are what microscopists call *weak-phase objects*. That phrase sounds technical, but its meaning can be grasped by thinking about a pane of nearly perfect glass. If you hold such a pane in front of a lamp, it does not cast a strong shadow. It does not absorb much light. Yet it can still change the light's *phase*—the timing of the wave crests—by slowing the wave slightly as it passes through. Your eyes are not very sensitive to phase on its own. You mostly notice brightness. So a phase-only object can be present and still appear almost invisible at perfect focus.

Electrons behave as waves too. A protein embedded in glassy water does not, in general, block many electrons outright. It nudges their wavefront. The information is real, but it is written in a code the detector cannot read directly: phase changes rather than simple darkening. In practice, the microscope must *convert* phase variations into intensity variations—into something that becomes visible as lighter and darker regions in the image.

Defocus, the deliberate choice to image slightly out of focus, is the classic way of doing that conversion. As was introduced earlier, defocus is not a blunder; it is a tactic. When the microscope is defocused, phase changes introduced by the specimen are transformed into contrast at the detector. The price is that defocus also distorts how different levels of detail are transferred. This distortion is described by a piece of mathematical machinery with a wonderfully concrete name: the *contrast transfer function*, or CTF.

The CTF is best imagined not as an equation but as a personality. It decides, for each size-scale of detail in the specimen, whether that detail will appear strongly, weakly, or even with its contrast inverted. It is frequency-dependent: broad features may be transferred one way, fine features another. It is also oscillatory: as you go to higher and higher spatial frequencies (finer detail), the CTF alternates between passing information with one sign and then flipping it. This is why two micrographs of the same specimen taken at different defocus values can have very different "textures." The microscope is not changing the specimen; it is changing the rules by which phase becomes brightness.

That oscillation has a curious consequence. Some bands of detail are effectively suppressed at a given defocus because the CTF happens to be near zero there. Those details are not necessarily absent from the electron wave; they are simply not expressed in the image. The practical response is to collect images at a range of defocus values, so that gaps in one image are filled by information from another. Then software estimates the CTF for each micrograph and corrects for it as part of the reconstruction process. This is one reason modern cryo work is inseparable from computation: the microscope does not hand you a straightforward photograph. It hands you a coded projection, and the code must be inferred and inverted.

All this is happening while the particles remain barely visible. That creates a human problem as well as a technical one. If you cannot easily see the particles, how do you focus reliably? How do you decide whether an area of the grid has the right thickness and cleanliness? How do you trust that an automated routine is selecting good targets rather than empty holes? In the early days of the field, this was part of the craft: experienced operators learned to read subtle cues in the ice and to interpret the grain of noise. A faint, slightly different mottling could suggest that macromolecules were present. The same cue could also be contamination. It could be a change in thickness. The microscope asked for judgment calls made on ambiguous evidence.

Two families of tools help make the evidence less ambiguous. One family throws away electrons that are more trouble than they are worth. The other changes the way phase information is turned into visible contrast, so that the particle stands out more clearly.

Start with the electrons you might prefer not to record. When an electron passes through the specimen, it can scatter elastically or inelastically. Elastic scattering changes the electron's direction without changing its energy; inelastic scattering involves an energy loss, often because the electron excites vibrations, plasmons, or other processes in the sample. Inelastic scattering is not a moral failing of an electron, but it does tend to blur the information you care about. The energy-lost electrons have been changed in ways that make them less tightly related to the specimen's structure at fine detail. They also contribute a diffuse background that washes out contrast.

An *energy filter* is the microscope's way of discriminating between these populations. It uses electron optics to spread electrons according to their energy and then selects a window, often favouring the "zero-loss" electrons that have not given up energy to the sample. The idea resembles choosing a narrow band on a radio: you tune out static and interference by selecting the part of the signal that best corresponds to what you want to hear.

The result can be a cleaner image with improved contrast, especially for thicker specimens where inelastic scattering is more common. Yet energy filtering is not simply a switch that makes everything better. It adds another complex optical element that must be aligned and kept stable. It can reduce the total number of electrons reaching the detector, because you are discarding part of the population, and under a strict dose budget any loss must be justified. It can also bring its own sensitivities and maintenance burdens. Still, the appeal is immediate to anyone who has stared at a low-contrast micrograph and wished for a way to separate whisper from hiss.

The second family of tools, *phase plates*, aims closer to the heart of the problem. If biological specimens encode much of their information as phase shifts, why not modify the microscope so that phase becomes intensity more directly, without relying so heavily on defocus?

This is the promise of the phase plate: an element placed in the back focal plane of the objective lens that changes the phase of the unscattered beam relative to the scattered beams. By introducing a controlled phase shift, the phase plate can turn what would have been an almost invisible phase object at near-focus into something with obvious contrast. In plain terms, it makes the particle stand out against the background without pushing the microscope far out of focus.

A prominent modern example is the *Volta phase plate*, associated with work by Radostin Danev and colleagues (2016) and further practical refinements with Baumeister's group (2017). The Volta phase plate exploits a phenomenon in which the electron beam itself conditions a thin film, producing a phase shift that can enhance contrast. One reason it attracted attention is that it promised to make small or delicate targets easier to see. If you are trying to image a relatively tiny protein complex that barely registers above the background, extra contrast is not cosmetic; it can determine whether particles can be picked reliably at all.

Phase plates also change the psychological experience of the microscope. With conventional defocus contrast, the best-looking images are often not the best for high-resolution information, and the image can feel perversely dependent on deliberately blurring the optics. A phase plate can produce micrographs that appear more "natural" to the eye: particles are clearer at lower defocus. This clarity can help focusing, targeting, and quality control. It can also make the field more accessible, because the data start to resemble what a newcomer expects an image to resemble.

But the phase plate is also temperamental. It introduces another surface inside a system that is already battling contamination. Any surface in the column can accumulate deposits over time, and a phase plate sits at a particularly sensitive location. If it changes as contamination builds, the phase shift it produces can drift. The phase plate can also be damaged or degraded by use. Practical workflows therefore involve monitoring how the phase shift behaves, deciding when a phase plate is "good," and sometimes living with a device that is powerful when it is behaving and maddening when it is not.

There is another, subtler point: even with a phase plate, a small amount of defocus can remain useful. That might sound like admitting defeat, but it is really about measurement. The CTF must still be estimated and corrected computationally, and defocus provides a signature that helps software infer the CTF robustly. Danev and Baumeister's practical work (2017) emphasized that phase-plate imaging is not simply "focus perfectly and forget the rest." The microscope is still encoding information through a transfer function; the encoding has changed shape, but it has not vanished. A small defocus can stabilize the downstream inference, like adding a slight accent to speech so a transcription algorithm can tell similar vowels apart.

This interplay between optics and computation can feel backwards at first. In everyday photography, the goal is to capture a faithful image directly. In cryo work, the goal is to capture an image that, together with a model of the microscope's behaviour, preserves as much recoverable information as possible. Sometimes that means deliberately introducing distortions that are mathematically tractable and then correcting them later. The art lies in choosing distortions that increase contrast without irreversibly erasing fine detail.

Consider a concrete case that has become common in the field: a dataset collected on a small protein where particles are barely visible in conventional micrographs. A phase plate can make particle picking feasible by increasing contrast at low defocus, reducing the temptation to push defocus so far that high-resolution information is damped and the CTF zeros become severe. Energy filtering can, in some setups, further clean the background, making the same particles easier to align. The combination can turn an ambiguous field of specks into a dataset with clear targets and more reliable alignment.

Yet every additional element also increases the number of ways the instrument can misbehave. An energy filter must be aligned; a phase plate must be conditioned and kept clean; both can be sensitive to the same enemies introduced earlier: contamination, drift, and the slow creep of instability. This is why cryo facilities often develop a repertoire of "modes" rather than a single best configuration. Some specimens benefit enormously from extra contrast tools; others are better served by the simplest optical path and a reliance on statistics from many particles.

The practical reality is that contrast is never free. It is always purchased with something: complexity, reduced throughput, stricter alignment demands, or an extra layer of modelling in software. The people running high-end instruments learn to think like signal engineers. They ask what fraction of the recorded electrons carry the information they need, what fraction are background, and how different choices move the balance. They also learn that the most fragile part of the system may not be a lens or a detector, but an assumption. If you assume a constant phase shift from a phase plate when it is actually drifting, the reconstruction will suffer. If you assume inelastic electrons do not matter when they dominate the background, you will overestimate how much detail is truly present.

Behind these technicalities sits a simple human desire: to be able to *see* what is otherwise too faint. The phrase "making shadows speak" is not just poetic. The images are, in a literal sense, shadows cast by scattering. The specimen does not announce itself with strong absorption; it whispers through phase. Energy filters help by quieting some of the background voices. Phase plates help by translating the whisper into a more audible register. Defocus, used carefully, provides a reliable grammar that software can parse.

When everything is working well, a micrograph can acquire a new kind of legibility. Not a dramatic change from grey to black-and-white, but an easing of uncertainty. Particles become more distinct from the ice. Alignments become more trustworthy. The reconstruction becomes less dependent on heroic amounts of data and more responsive to the information already present in each exposure. The microscope has not made the specimen less delicate; it has simply become better at listening.

3.4 From Solitary Art to Molecular Factory

For a long time, a productive day at a cryo microscope could be described in the same tone people use for a day of fly-fishing: patience, a little luck, and a practiced feel for subtle conditions. The operator would sit at the console, find a promising square on the grid, hunt for holes with suitable ice, adjust focus, check astigmatism, expose a frame, and then begin again. Each decision was embodied. You learned the instrument the way a musician learns a particular violin: by touch, by sound, by the way it behaved on damp days. A good dataset depended on that intimacy, and the scarcity of images was part of the culture.

Then the pace changed. Direct detectors began producing not just images but movies, and the value of a session came to be measured in thousands of exposures rather than dozens. That shift created a pressure that was almost industrial: if the camera could record continuously, why should the microscope spend its nights idle? Why should an expensive instrument, built to hold stability on the scale of atoms, be limited by the endurance of the person in the room?

Automation entered cryo work partly as an answer to that economic question, but it also answered a scientific one. When data are scarce, it is tempting to treat each micrograph as an object of individual attention. When data become abundant, individual attention becomes a bottleneck, and the workflow must be designed so that routine decisions are made consistently. The goal of automation is therefore not merely speed. It is *standardization*: the ability to collect large datasets under controlled conditions, with enough metadata to understand what happened later.

Two software systems are often named in this context because they captured the practical philosophy early and shaped how facilities operate: Leginon and SerialEM. Both use calibration and image-based targeting to select suitable grid holes and record acquisition parameters systematically. Instead of the operator manually steering the stage and deciding where to expose, the software can map a grid square, identify holes, rank them according to criteria such as apparent ice thickness and cleanliness, and then move through them in a disciplined pattern. The microscope becomes, in effect, a robot guided by images.

That does not mean the human disappears. It means the human moves up a level, from steering every movement to designing and supervising the strategy. An operator might still decide which grid squares are worth exploring, what defocus range is appropriate for a given sample, and how strict the targeting criteria should be. But once the strategy is set, the machine can execute it with a steadiness that does not fatigue at 2 a.m.

The most visible mechanical sign of this change is the autoloader. Instead of inserting a single holder by hand, many modern instruments store multiple grids in a cryogenic cassette and use a robotic mechanism to load them into the column. The mundane benefit is convenience: you can screen several grids in one session without repeatedly breaking vacuum. The deeper benefit is reproducibility and cleanliness. Every manual insertion is a moment when frost, dust, or hydrocarbons can be introduced. Robotic loading through controlled vacuum locks reduces those risks and reduces variability between sessions.

Screening itself, once a highly personal activity, has also become more systematic. A typical workflow now begins with a low-magnification atlas: a stitched map of the entire grid. From this map the software chooses squares that appear intact and not overly contaminated. Then it takes intermediate-magnification images of those squares to find holes. From those hole images it selects targets for high-magnification exposures. Each step is based on the previous step's image, so the targeting is literally image-guided.

At each stage, the software can store not only the images but the context: stage coordinates, lens settings, defocus targets, dose rates, time stamps, and identification of which grid and which hole produced which movie. This database-backed metadata is one of the quiet foundations of modern cryo work. Without it, a dataset of ten thousand movies is just a pile of files. With it, the dataset becomes traceable. You can ask questions later that would have been impossible in the film era: did image quality drift over the night as contamination slowly accumulated? Are certain grid squares systematically worse? Did the microscope's astigmatism change after a refill? Does one region of the detector behave differently? Automation does not just record more; it records *with memory*.

A second pillar of the factory-like workflow is that acquisition is chained to preprocessing and quality control. The raw output of a direct detector is a stack of frames, and those frames are typically motion-corrected almost immediately. MotionCor2, widely integrated into preprocessing pipelines, can align frames and correct for local motion, producing a sharper summed image and metrics describing how much movement occurred. Then CTF estimation is run on the motion-corrected micrograph to assess whether the image has plausible optical parameters and whether the high-resolution information is present.

These early steps are standardized enough that they can be treated as routine, not as artisanal choices. In many facilities, by the time an operator returns the next morning, the night's movies have already been aligned, CTF parameters estimated, and basic quality plots generated. What used to be a day's worth of manual inspection becomes a dashboard. A dataset can be triaged quickly: keep collecting, adjust settings, or stop because the grid is not worth the time.

This is where automation begins to change scientific behaviour. On-the-fly analysis shortens feedback loops. In the older style, you might collect for hours before realizing that the drift was too high, that contamination was creeping in, or that the ice was systematically too thick. Now those problems can be flagged after the first few dozen movies. The microscope becomes a system that can correct its own course, at least partially, rather than a machine that must be driven continuously by intuition.

Even particle picking, once done by hand with a kind of connoisseur's attention, has been pulled toward automation. When particles are plentiful and obvious, automated picking is straightforward. When particles are small or the background is messy, automated picking can fail in ways that are hard to notice unless you check. A common compromise is semi-automation: the software proposes picks, and the human inspects a subset for plausibility. The key change is again one of scale. The question is no longer, "Can I pick these particles perfectly?" It becomes, "Can I pick them consistently enough that errors average out, and can I detect systematic bias early?"

The transformation can be illustrated with a simple vignette that is now routine in many national and regional facilities. A researcher arrives with a handful of grids prepared over the previous week, each representing days of biochemical work. They may have only a narrow time slot on a shared instrument, and they may be travelling. The facility staff load multiple grids into an autoloader, then run a screening script: atlas, square selection, hole finding, a small number of test exposures per grid. The software ranks the grids by ice quality and preliminary image metrics. Within hours, the team knows which grid is worth an overnight collection and which should be abandoned or remade. In earlier decades, that decision might have required repeated sessions and a great deal of subjective judgment. Now it can be made with data.

The emotional impact is not trivial. When microscope time costs thousands per day and sample preparation may have taken months, the ability to fail quickly is a kind of mercy. It prevents the slow waste of collecting beautiful-looking images that later turn out to be unusable. It also changes how people learn. Instead of apprenticeship by osmosis at the console, training can include reading quality metrics, understanding failure modes, and learning to diagnose problems from the output of standardized pipelines. Skill still matters, but it is expressed differently: as the ability to interpret a system's behaviour rather than to perform every action directly.

That said, the factory metaphor has limits. A factory implies uniform raw material. Cryo specimens are not uniform. Two grids made from the same protein can behave differently depending on subtle differences in preparation: the distribution of particles, the thickness of ice, preferred orientations, the presence of aggregates, or the tendency of particles to stick to the air-water interface. Automation can consistently collect images of whatever is there, but it cannot guarantee that what is there is scientifically useful.

This tension shapes how automation is designed. Targeting algorithms can use proxies for ice thickness and contamination, but those proxies are imperfect. A hole that appears ideal at one magnification may contain very few particles. Another may contain particles but also a gradient in thickness that complicates processing. Some samples demand unusual conditions: very low defocus, phase plates, energy filters, tilted data collection to overcome orientation bias. Each deviation from the standard pipeline increases the chance that an automated script will make an inappropriate assumption.

So the "molecular factory" is better thought of as a flexible production line staffed by vigilant supervisors. The system can run for hours without intervention, but it thrives on periodic checks: is the motion increasing as the stage warms? Are the CTF fits degrading? Is the rate of good particle picks dropping? Is contamination creeping in as frost accumulates somewhere in the transfer path? These are not purely technical questions. They are questions about resource allocation and trust. At what point do you stop collecting and try a different grid? At what point do you accept that the microscope is fine and the sample is the limiting factor? Automation provides numbers, but it does not make those decisions painless.

Facility-scale considerations become part of the science. A single instrument is not just a tool; it is a shared resource that must be kept healthy. Contamination control is no longer the private problem of one graduate student; it is a communal responsibility. Alignment monitoring becomes routine. Training becomes formalized because mistakes are costly not only for one project but for everyone using the instrument. Thompson and Iadanza (2019) emphasized these facility realities when discussing the establishment of national cryo capabilities: staffing, workflow, infrastructure stability, and disciplined practices are as crucial as the hardware itself.

The data firehose created by automation brings its own kind of fragility. Storing tens of thousands of movies is not merely a matter of buying disks. It requires an organized data model. Files must be named and linked to metadata. Processing must be reproducible so that if a dataset is reprocessed months later, the steps and parameters are known. Transfer speeds matter because a slow network can become the bottleneck that the microscope used to be. Backup policies matter because losing raw movies can mean losing the chance to revisit a question with improved algorithms.

In the film era, the artefacts of work were tangible: boxes of negatives, notebooks, stacks of prints. In the automated era, the artefacts are databases, logs, and directory trees that can be copied and shared but also easily scrambled. This has changed the social structure of projects. More people can collaborate, because data can be shipped electronically. More expertise becomes specialized, because one person may focus on sample preparation while another focuses on computational pipelines. At the same time, the boundary between "data collection" and "data analysis" has blurred. The choice of acquisition parameters is now made with an eye to downstream processing: motion correction, CTF estimation, particle picking, classification. A dataset is planned as an end-to-end object.

Perhaps the most striking symbol of this change is what happens overnight. In many labs, the microscope collects while the building sleeps. The detector records movies, the software targets holes and maintains focus, the pipeline corrects motion and estimates the CTF as files appear on disk. In the morning, the researcher does not begin by developing film or scanning negatives. They begin by looking at statistics: distributions of drift, defocus, and resolution estimates; a gallery of representative micrographs; early two-dimensional class averages. The craft has not vanished, but it has been reorganized. Where a microscopist once demonstrated skill by coaxing a few perfect images from an unwilling specimen, they now demonstrate skill by designing a process that can produce thousands of consistently useful images, and by knowing when the numbers are lying.

A molecular factory, at its best, does not feel cold or impersonal. It feels like a hard-won relief from scarcity. It allows a researcher to stop gambling on single exposures and start thinking statistically. It makes it practical to screen multiple conditions, compare grids, and iterate on sample preparation with real feedback. It also makes the microscope's silence more productive. When the room is stable and the machine is aligned, the system can simply work, collecting its quiet torrents of data, each movie a tiny record of electrons counted and shadows translated into structure.

Chapter 4

From Snowy Snapshots to Solid Shapes

At first glance, a cryo-EM micrograph looks less like a portrait than a snowstorm: faint specks buried in damage, drift, and statistical uncertainty. Yet from these nearly unreadable glimpses, researchers learned to recover stable molecular shapes, separate one conformation from another, and decide when a map deserved belief rather than hope.

4.1 A Face in the Snow

A raw cryo-electron micrograph has an awkward personality. At first glance it looks like weather: a grey field peppered with specks, ripples, stains, and the occasional sharp edge where a piece of carbon support intrudes. Somewhere in that wintry static sit the molecules you cared enough to purify, buffer, and freeze. Yet if you point to any single speck and declare *there it is*, even an expert will hesitate. One particle image rarely has the courtesy to look like anything. The paradox at the heart of single-particle cryo-EM is that no individual snapshot is good enough on its own, and the method succeeds precisely by refusing to bet on any one of them.

The refusal is not philosophical; it is forced by physics. The electron dose must be kept low or the specimen is altered while you are still looking at it. Low dose means low contrast, and low contrast means that the signal from a protein or ribosome arrives buried under shot noise, detector noise, and the texture of vitreous ice. The microscope records an honest image, but honesty is not the same as clarity. Single-particle analysis is the decision to treat that murky image not as a failure but as a census: a record containing many faint copies of the same object, each copy offering a slightly different view.

That idea sounds almost too convenient. Where do all these copies come from? From the way the sample was prepared. A well-behaved grid contains countless molecules scattered through the ice, each one trapped in a random orientation as the water vitrified. In everyday life randomness is the enemy of measurement; in cryo-EM, random orientation is a gift. If every particle lay the same way, you would learn only one face of the molecule, no matter how many images you took. But if the particles are jumbled like coins tossed onto a table, the dataset naturally contains many different projections. Each particle is a two-dimensional view of one three-dimensional object, and a collection of such views can, in principle, be combined to infer the object's shape.

The word *projection* matters, because it quietly states the bargain. When the beam passes through a thin specimen, the image on the detector can be treated, to a good approximation, as a projection of the specimen's electron-scattering potential along the beam direction. It is like a shadow, except that it is made from electrons rather than light, and the image contrast is shaped by the microscope's optics in ways that can invert and blur features. But the essential geometry holds: every time a particle sits at a different orientation, it produces a different projection. A sphere looks like a circle from every direction; a hammer looks like a line from some directions and a complicated silhouette from others. The more varied the views, the more constrained the reconstruction.

None of that helps unless you can first locate the particles. This is where cryo-EM can feel strangely like looking for faces in a snowstorm. A micrograph contains tens to thousands of particles, and those particles can be only a few dozen pixels across. On a good day you see faint, consistent shapes: the donut of a chaperonin, the boxy outline of a ribosome, the spikes of a virus. On many days the particles are close to invisible. The task called *particle picking* is simply deciding, for every micrograph, which small patches contain a particle worth keeping.

Early on, particle picking was literal: a researcher would stare at the micrograph and click, click, click, guided by experience and a sense of what "real" looks like under that particular ice thickness and defocus. It was tedious, but it also trained a kind of visual literacy. People learned, in their fingertips, what ice contamination looks like, what aggregated protein looks like, what a broken carbon edge does to contrast, and how a good particle does not need to be obvious to be usable. That literacy still matters, even if the clicking has largely been delegated to software. Modern pickers can use templates, neural networks, or heuristics tuned to blobs and edges; whatever their internal logic, they are trying to solve the same problem: box out many candidate particles while missing as few true ones as possible.

The "box" is not a metaphor. Each picked particle is cut from the micrograph into a square image, typically just large enough to include the particle and a margin of surrounding background. This creates a stack: thousands to millions of little squares, each one centered (one hopes) on a single molecule. If the micrograph is a landscape, the particle stack is a shoebox of postage stamps.

And at this point the method makes its second daring move: it treats those stamps as if they were comparable. The raw stamps are not aligned; one particle is shifted left, another right; some are rotated; some include a particle that is partially outside the box; some contain no particle at all. Even among true particles, contrast varies with local ice thickness, the particle's precise height within the ice, and the optics of the microscope at that moment. The immediate goal is not to build a 3D model. It is to find commonality.

Averaging is the most intuitive way to find commonality, because it uses a simple statistical fact: if you add many independent noise patterns together, the noise tends to cancel, while the consistent part adds up. Imagine photographing the same dim object in a dark room with a jittery camera sensor. Each photo is grainy, and the object is barely visible. If you could perfectly align the photos so the object sits in the same place each time, then averaging them would reduce the grain. The object would emerge. In cryo-EM, the object is a particle projection, the grain is a mix of real noise and background, and alignment is the hard part.

Alignment means estimating how each particle image should be shifted and rotated to match a reference. But where does a reference come from when nothing is visible? The earliest strategies were pragmatic: start with a rough reference (perhaps even a blurred average), align everything to it, compute a new average, and repeat. Iteration is powerful, but it has a dark side: if the reference contains artifacts, the algorithm can persuade the data to imitate them. That is why single-particle analysis has always balanced ambition with caution, and why it became increasingly statistical as the field matured.

A practical way around the lack of an obvious reference is to avoid pretending that all particles look the same. They do not. Even if the molecule is perfectly rigid, its projection changes with orientation.

A side view of a ribosome and a top view of a ribosome are both correct, but averaging them together produces nonsense: features smear into fog. The solution is to group particles into sets that represent similar views and only then average within each set. The term *class average* captures the idea. Each class is a pile of particles that appear to be the same projection up to shift and rotation. Averaging within the class boosts the signal for that particular view. Instead of one indecipherable stamp, you get a clearer portrait: edges sharpen, cavities appear, and symmetry—if the particle has it—begins to announce itself.

There is a humanly satisfying moment here, familiar to many cryo-EM practitioners: the first time a dataset yields recognisable class averages. Until then, everything feels like faith. The micrographs are ugly; the picked particles look like static; skeptics in your head whisper that you have frozen buffer, not biology. Then the averages appear—rings, blades, lobes—shapes that are too consistent to be accidental. It can feel like watching a face emerge in a developing photograph, except that the "chemistry" is computation and the face is a molecule you may have studied for years without ever seeing.

The logic behind class averaging also hints at why random orientations are so valuable. Each class average is, in effect, one view of the molecule. A well-sampled dataset contains many different class averages corresponding to different orientations. If you could somehow arrange them on a sphere—each one tagged with the viewing direction it represents—you would possess the raw ingredients for a 3D reconstruction.

The leap from many 2D projections to one 3D object is an old problem in imaging. Medical CT scanners solve it by rotating an X-ray source and detector around a patient, collecting projections at known angles, then reconstructing a 3D volume by a mathematical inversion. Cryo-EM does not rotate the molecule on demand;

instead, it relies on the statistical gift of random orientation. That means the angles are unknown, and determining them becomes part of the inference. The reconstruction problem is therefore entangled: to build a 3D map you need orientations, but to find orientations you need a 3D map.

Historically, this entanglement was one reason single-particle analysis took time to mature. Joachim Frank was among the crucial figures in turning the idea into a workable discipline, developing approaches to alignment and classification that could wrestle signal from noise and assign images to consistent views. The conceptual shift was subtle but profound: instead of treating noisy images as flawed measurements of a known object, treat them as samples from a distribution of unknown views, and use statistics to infer the underlying structure. That shift, developed over decades and refined by many groups, made it possible to speak sensibly about reconstructing asymmetric, flexible biological machines rather than only the most regular, crystalline specimens.

Even without diving into equations, one can understand the geometry with a physical analogy. Suppose you are handed thousands of torn scraps from the same photograph, each scrap showing a small region at a different rotation and with different blur. If you can cluster similar scraps together and average them, you recover clearer patches. If you can then figure out how the patches relate—what borders match what—you can rebuild the photograph. Single-particle cryo-EM is doing something analogous, except that the "scraps" are not regions but projections, and the matching is done in Fourier space, where projections correspond to slices through the 3D transform of the object. The mathematics is beautiful, but the emotional truth is simpler: repeated, consistent hints add up to certainty.

Of course, the method's promise rests on a major assumption:

that the particles you are averaging really are the "same". Sometimes they are not. Samples can contain a mixture of intact complexes and broken subassemblies, different binding partners, aggregated junk, or the same complex caught in different conformations. Even a perfectly homogeneous sample can appear heterogeneous because the ice itself introduces variation—different background textures, different thickness, different charging behaviour under the beam. Particle picking will include contaminants; alignment will occasionally lock onto noise; averaging will sometimes invent crispness where none exists. Single-particle analysis therefore lives by a rule that sounds backward until you have tried it: be generous early, then ruthless. Pick a lot, accept uncertainty, and only later decide what deserves to contribute to the final portrait.

That rule has a practical reason. If you pick too conservatively, you risk throwing away rare but essential views. A molecule may have preferred orientations on the grid, lying "face down" most of the time. In that case, side views may be scarce—perhaps only a few percent of the particles—and they might look especially noisy because they are less common and harder to recognise. Yet those rare side views can be the difference between a reconstruction that is a flattened impression and one that is truly three-dimensional. In a sense, the method asks you to be patient with the crowd, because the crowd contains the missing angles.

Another reason to be generous early is that alignment itself becomes easier as the signal improves. There is a bootstrap effect. Rough class averages, even if imperfect, are better references than raw noise. Better references enable better alignment, which improves the averages, which improves the references again. This is why so much of cryo-EM processing feels like coaxing. The data are reluctant at first, and then, once guided into the right coordinate system, they become unexpectedly eloquent.

One can also see why an individual particle image is both precious and nearly useless. It is precious because it is a unique physical event: electrons scattered by one molecule, frozen in one orientation, at one moment, under one set of microscope conditions. It is nearly useless because the event is recorded under such tight dose limits that its details are submerged. Cryo-EM's triumph is to treat each event as a vote rather than a verdict. No single vote decides the outcome; collectively, they reveal a stable pattern that persists across noise, across micrographs, and across time.

There is a quiet ethical lesson in that technical strategy. Biologists often speak of molecules as if they were clean, isolated, and fully knowable, but inside cells they are crowded, jostled, and constantly interacting. Even in a purified sample, the molecule is not an idealised diagram; it is an object with variability, with surface chemistry, with a history of how it met the grid and the ice. Single-particle cryo-EM does not deny that messiness. It makes a disciplined bargain with it: accept the messiness at the level of individual images, then use large numbers and careful grouping to recover the reproducible core.

By the time a dataset yields strong class averages, something has already happened that is easy to overlook. You have converted an image that looked like weather into a set of views that look like an object. The object is still anonymous—you do not yet know its full shape—but it has ceased to be a rumour. You can point to a class average and say: this is a top view, this is a side view, this is a tilted view; these edges recur; this cavity stays put. In other words, you have found a face in the snow, not because the snow stopped being noisy, but because repetition made the face undeniable.

4.2 When the Camera Learns to Watch a Wobble

A long exposure has a kind of dignity. It implies patience: open the shutter, let the light accumulate, and trust that the scene will hold still long enough for detail to build. Early cryo-EM treated electrons the same way. The microscope collected an exposure, the detector produced a single image, and whatever happened during that fraction of a second was baked into the result. If the specimen drifted, if the stage crept, if the ice flexed under the beam, the picture simply became softer. Blurring was accepted as a fact of life, like a hand-held photograph taken in dim light.

The uncomfortable twist is that cryo-EM is dim by necessity. The beam must be gentle, which means the image begins life starved of contrast. Any extra blur is not a small insult; it can be the difference between seeing a crisp edge and seeing nothing at all. The most valuable details are the finest ones, and the finest ones are the easiest to smear away. A protein's overall silhouette can survive a slight wobble. The subtle corrugation that hints at a helix, or the crisp corner of a beta sheet, cannot.

The breakthrough that changed the mood of the field was not simply "better cameras" in the everyday sense. It was the realisation that an exposure did not have to be a single still. If the detector could record a rapid sequence of frames during the same total dose, then motion would stop being an invisible saboteur. It would become something you could measure. And if you could measure it, you could reverse it, at least partly, by shifting the frames back into register before you ever tried to align particles to one another.

This is where direct electron detectors enter the story with a kind of quiet drama. Traditional detectors relied on a conversion step:

electrons struck a scintillator, producing photons, which were then detected by a camera sensor. The conversion spread out the signal, like ink bleeding in paper, and it cost sensitivity. Direct detectors register electrons more directly, with far better efficiency, and crucially they can read out quickly enough to record many frames per second. Li and colleagues (2013) showed how electron counting and frame-based correction could pull back detail that had been slipping through the fingers of the old workflow. What had been a single blurred photograph could be replaced by a short movie, and that movie contained a record of the specimen's restlessness.

The term *dose fractionation* sounds technical, but the idea is homely: if you have a limited budget of electrons, you can spend it all at once or in small instalments. Spend it all at once, and you get one image whose blur is the sum of whatever motion occurred. Spend it in instalments—many subframes whose total equals the same overall dose—and you can ask how the scene shifted from one instalment to the next. Importantly, this does not violate the dose limit. The specimen receives the same total damage. What changes is that the camera now tells you *when* each electron arrived, not merely *where* it landed.

That timing matters because beam-driven motion is not necessarily smooth. Even without revisiting the causes in detail, the behaviour can be sketched in familiar terms. Some movement is global: the whole field of view glides a little because the stage is not perfectly rigid. Some movement is local: different regions of the ice bend or relax differently, so particles at one corner drift in one direction while particles elsewhere drift in another. Even within a single micrograph, the motion can be patchy, as if the ice were a trampoline flexing under a light tap.

The first generation of movie alignment treated the problem as if the entire image were a rigid sheet: estimate a single shift for each frame, apply it, and sum the corrected frames into a sharper picture. That alone could be transformative. A summed exposure that previously looked "fine" could suddenly reveal stronger, more regular Thon rings in its power spectrum—the telltale ripples that signal coherent contrast transfer and that practitioners use to judge whether the image contains reliable fine detail. Those rings are not merely a diagnostic decoration; they are a sign that the microscope and the specimen have preserved the subtle oscillations in contrast that carry the smallest features. When alignment tightens the rings, it is a visual cue that information has been recovered from what would otherwise have been blur.

But global alignment has a ceiling, because the specimen is not a rigid sheet. If you correct the average drift, you may still leave behind a residual smear: particles in the top-left corner that drifted more than the mean will remain slightly blurred, and particles that drifted less may become over-corrected. The natural next step is to let different parts of the frame have different motion trajectories. This is what people mean when they speak of *anisotropic* motion: not simply "moving," but moving differently in different directions and places. Zheng and Palovcak and colleagues' MotionCor2 (2017) is an emblem of that shift, offering a practical way to model local, non-uniform motion across the image rather than forcing all particles to share the same correction.

It is tempting to imagine that motion correction works like stabilising shaky video on a smartphone: compute the jitter, cancel it, and the world stands still. Cryo-EM is harsher. The movie frames are extremely noisy, each one receiving only a small fraction of the already-limited dose. Aligning them is like trying to stabilise a night-time video where each frame is almost black. The reason it works at all is that the motion is usually coherent enough to be inferred statistically, and because the field of view contains enough structure—edges, textures, repeated patterns—that shifts can be detected even when individual particles are faint.

That faintness creates a subtle but important practical change in attitude. Under the old still-image approach, the micrograph was the unit of truth. If it looked blurry, it was blurry; if it looked sharp, it was sharp. Movie-mode turns the micrograph into an object with internal time. Sharpness becomes something you can redistribute: recover some crispness from the early part of the exposure by correcting its motion, and choose how much to trust the later part of the exposure once radiation damage has started to bite.

This brings dose weighting into view. Even if all frames were perfectly aligned, they would not all be equally informative. The specimen changes as it is irradiated. Fine structural features are the first to suffer; broad outlines persist longer. That means the earliest frames, in principle, carry the most delicate detail, because the molecule is closest to its original state. Yet those earliest frames can be the most motion-affected, because the specimen may "settle" under the initial impact of the beam. Later frames may be steadier but more damaged. Cryo-EM processing therefore faces a trade-off that would sound absurd in ordinary photography: the clearest frames are not necessarily the most useful, and the most useful frames are not necessarily the clearest.

Dose weighting is the computational answer to that trade-off. Instead of summing all frames equally, you assign different weights to different parts of the movie, typically giving more influence to earlier, less-damaged frames when it comes to fine detail, while still using later frames for the lower-frequency information that survives damage. The point is not to rescue the specimen from harm—the harm is already done—but to make the best use of what each moment of the exposure can honestly contribute.

A concrete way to feel this is to imagine drawing a portrait while the sitter slowly changes expression. The first few seconds capture the authentic face but your hand is unsteady; later your hand steadies but the sitter begins to grimace. If you average all your sketches without thinking, you get a blurred face with a vague, wrong mouth. If you could align the sketches and decide, pencil stroke by pencil stroke, which moments best represent the features you care about, you might recover a faithful portrait. Dose weighting is a version of that selectivity, enacted in Fourier space and guided by models of damage.

The impact of movie-mode is often felt most sharply when comparing datasets that, to the naked eye, seem equally unimpressive. A newcomer might look at two micrographs—one recorded as a still, one as a movie later aligned and weighted—and shrug: both are grey, both are noisy, both show little dots. Yet downstream, the movie-derived micrograph produces markedly crisper class averages and a reconstruction with tighter features. This disconnect between what the human eye perceives and what the algorithms can extract is one reason the "resolution revolution" could feel magical from the outside. It was not magic; it was the recovery of information that had been present but scrambled over time.

The historical anecdote here is less about a single dramatic instrument unveiling and more about a change in what counted as a solvable problem. Before direct detectors, high ambitions often required heroic sample behaviour: exceptionally rigid particles, strong signal, fortunate orientations, and an unusual tolerance for the compromises of low-dose imaging. After direct detectors and movie alignment, many more samples became tractable. Not easy—never easy—but tractable. It became plausible to treat beam-induced motion as a correctable nuisance rather than a fixed ceiling. That shift altered not only the attainable resolution but also the daily psychology of the field. When a map looked soft, you no longer had to assume the biology was hopelessly flexible or the microscope misaligned. Sometimes the culprit was simply that the camera had been asked to average a wobble it had never been allowed to observe.

There is also a deeper consequence that reaches beyond sharpness. Movie-mode changes what "one image" means. In the still era, downstream processing began with a micrograph and proceeded to particle picking and averaging, as though the micrograph were a static document. In the movie era, the first act of interpretation is temporal: deciding how to combine frames, how to model local drift, how to weight damage. The reconstruction's credibility begins earlier than many people expect, in decisions made before a single particle has been aligned to a reference. A beautiful final map can be nudged upward or downward in quality by a choice made at the very start about how to treat time.

This early-step importance explains why motion handling has grown steadily more sophisticated. The specimen's movement is not only a property of the microscope stage; it is an emergent behaviour of water, carbon support, charging, and the mechanical stresses of cooling and irradiation. Different grids can behave differently. Even different holes in the same grid can behave differently. Local motion correction methods acknowledge that reality by letting the correction be flexible: a compromise between assuming everything moves together and assuming everything moves independently.

It also explains why movie-mode is not simply "more data." It is better-structured data. A single still image can be thought of as the sum of many sub-exposures, but it offers no way to disentangle them. A movie makes that sum explicit. It turns blur from an irreversible averaging into a reversible misalignment, at least within limits. That reversibility is what allows the faintest features to survive long enough to be reinforced by the crowd later on.

At the end of this chain of ideas sits a surprisingly ordinary object: a sharper micrograph. Not sharp in the glossy way of a magazine photograph, but sharp in a very specific, hard-won sense. Edges are slightly more coherent. Background texture is slightly less smeared. The Fourier space diagnostics look healthier. And most importantly, when the particle stack is built from these corrected images, the averages stop fighting the blur imposed by the camera's ignorance of time. The computational effort that follows—alignment, classification, reconstruction—starts from a footing that is less slippery.

The most poignant part of the story is that the specimen never truly holds still. Even in the best-corrected movies, there is always residual motion, always some damage, always a limit set by physics and by the particular fragility of the sample. The achievement of movie-mode is not the elimination of wobble, but the decision to treat wobble as measurable behaviour rather than fate. Once the camera learned to watch, the microscope stopped pretending that an exposure was a single moment, and began admitting that seeing is something that happens over time.

4.3 Teaching the Crowd to Sort Itself

A cryo-EM dataset begins as a promise and a mess. After picking, boxing, and the first rounds of averaging, there is usually a rush of relief: something real is in there. But relief can be misleading. A particle stack is not a gallery of identical portraits; it is closer to a lost-and-found box. Alongside the molecules you wanted are molecules you did not: broken complexes missing a subunit, dimers that formed in the freezer overnight, bits of carbon edge, specks of ice contamination, aggregates, and particles caught at awkward angles. Even if the sample is chemically pure, the biology can still be mixed in a deeper sense: a machine may open and close, bind and unbind, twist and relax. Averages are generous; they will faithfully blend all of these into a smooth, persuasive blur.

That blur is not just a cosmetic flaw. It can hide the very thing you came for. Imagine trying to photograph a friend who cannot decide whether to smile or frown. If you overlay a hundred shots, you will obtain a face with softened cheeks and a mouth that belongs to no human expression. The image might even look "clean" because the randomness averages away, yet it is wrong in a more fundamental way: it does not correspond to any single state that ever existed. Cryo-EM faces the same problem, except the expressions are molecular: a ribosome rotating between functional states, an ion channel opening and closing, a viral spike protein breathing between conformations. Without sorting, the map becomes a compromise that satisfies no state.

The field's answer is classification: letting the dataset organise itself into piles whose members truly resemble one another. The trick is doing this when each individual particle image is still noisy and ambiguous. You cannot simply measure a few features with a ruler and cluster by eye. Classification has to be statistical, tolerant of uncertainty, and willing to revise its own decisions as the underlying picture improves.

A good place to start is with an unromantic question: which images should not be in the dataset at all? Particle picking is designed to err on the side of inclusion, and the result is that the stack contains rubbish. Some boxes contain no particle; others contain particles at the edge, double-particles, crystalline ice, or thick regions where the background overwhelms the signal. If these are allowed to participate fully in averaging, they do not merely "add noise." They can pull averages toward spurious shapes, distort alignment, and waste computational effort. The first sorting step, usually called *2D classification*, is a way to ask the data what recurring views exist and which images refuse to belong to any view.

The output of 2D classification can be surprisingly emotional for practitioners because it feels like meeting the sample for the first time. One class average is a crisp, recognisable projection: perhaps a ring, a lopsided wedge, a symmetric barrel. Another is a smudge. Another is a featureless blob. Another looks like a piece of carbon edge. Even without any mathematics, you can see the logic: keep the classes that look like plausible projections of a molecule; discard the classes that look like background or nonsense.

Yet it would be a mistake to think of 2D classification as mere housekeeping. It also reveals the *view distribution*: which orientations are common and which are rare. This matters because 3D reconstruction is constrained not only by the number of particles but by the diversity of angles they represent. A dataset dominated by one view might give beautiful class averages and still struggle to yield a well-defined 3D shape, much as a sculptor cannot carve a head from photographs taken only from the front. Classification provides a census of perspectives.

Once the dataset has been cleaned, the reconstruction problem returns with its old circularity: to assign a particle an orientation you need a 3D model; to build a 3D model you need orientations. The way out is an *initial model*, a first guess at the object that is good enough to begin iterative refinement. "Good enough" is a modest standard. The initial model need not contain fine detail; in fact, it should not. An initial model is safest when it captures only the broad shape, so it guides orientation assignment without imposing an imagined atomic story.

Historically, generating an initial model was a point where the method could stumble. If you began with an incorrect reference, refinement could be seduced into confirming it. This is the uncomfortable phenomenon of model bias: when an algorithm keeps aligning noisy images to a reference, it can begin to pull the reconstruction toward the reference's features, even if those features are not supported by the data. The solution has been partly cultural (a suspicion of overly detailed starting references) and partly algorithmic (methods that can generate initial models more ab initio, with weaker assumptions).

The emergence of RELION, introduced by Sjors Scheres (2012), marked a turning point in how the community talked about these problems. RELION framed single-particle reconstruction as Bayesian inference, a way of combining the evidence in the noisy images with reasonable priors and regularisation. "Bayesian" can sound like a mathematical badge, but the practical effect was very concrete: refinement and classification could be done in a way that was less eager to over-interpret noise. The program's popularity also helped standardise what had previously been a patchwork of lab-specific practices. A dataset could be processed with a workflow that many other groups understood, critiqued, and improved.

Around the same time, improvements in computational speed changed what was psychologically possible. Classification is computationally hungry because it asks, repeatedly, whether each particle fits better in one class or another and at what orientation. Doing this for hundreds of thousands of particles was once a grim overnight endeavour. With modern GPUs and smarter algorithms, it became something you could iterate on during a project, not merely endure at the end. That shift encourages exploration: trying different class numbers, different masks, different subsets of particles, looking for hidden states rather than settling for the first map that appears.

cryoSPARC, introduced by Ali Punjani and colleagues (2017), exemplified this acceleration and a slightly different philosophy. Its strength lay in rapid, largely unsupervised ways to generate initial models and to perform heterogeneous reconstruction. In everyday terms, it helped researchers ask “what shapes are actually in my data?” early, before they had committed to a single reference. That early glimpse can prevent months of work from being invested in a consensus map that quietly averages multiple states into mediocrity.

Once an initial 3D model exists, refinement becomes a kind of negotiation between the model and the images. Each particle is assigned an orientation and, often, other parameters describing how it should be interpreted under the microscope’s optics. Using those assignments, a new 3D map is computed. The improved map then serves as the reference for another round of orientation assignment. This cycle repeats until the map stops changing appreciably. The appealing part is that the map and the particle parameters improve together, each making the other easier to estimate.

The dangerous part is that the cycle can also reinforce mistakes. If a subset of particles is consistently misaligned—perhaps because they are junk, perhaps because they represent a different conformation—then refinement may try to accommodate them by blurring the map or by inventing compromise features. The map becomes a diplomat, smoothing over disagreements rather than insisting on truth. Classification is the way to refuse that diplomacy.

3D classification generalises the idea of 2D classification into three dimensions. Instead of grouping images by similar 2D appearance alone, particles are grouped by which 3D map they best support.

Conceptually, you are asking: does this particle fit better with a map where the domain is open or closed? With a map where a ligand is bound or absent? With a map where a subunit is present or missing? Practically, the algorithm maintains several candidate 3D maps at once and assigns each particle, probabilistically, to the map it most likely came from, given the noise.

A thought experiment makes the necessity vivid. Consider a protein with a mobile "lid" domain that can be either open or closed. Suppose the two states are equally populated. If you refine all particles together into one consensus map, the lid region will appear fuzzy, perhaps interpreted as flexibility or poor resolution. Yet in reality the lid may be perfectly ordered in each state. If you split the dataset into two classes and refine them separately, each class map can show a crisp lid, one open and one closed. Nothing "improved" in the microscope; the improvement came from no longer forcing incompatible images into the same average.

This logic becomes even more important for large molecular machines that step through functional cycles. Ribosomes, for example, are not statues; they are factories with moving parts. Many of the most influential cryo-EM studies in the past decade have not simply produced a single ribosome structure, but a series of state-specific maps that together suggest a mechanism. The achievement is not only resolution but narrative: being able to say, with evidence, that a particular domain rotates at a particular step, or that a tRNA occupies distinct sites in sequence. Without sorting, those mechanistic stories dissolve into blur.

Sorting does not only separate dramatic conformational switches. It can also identify compositional mixtures. A sample might contain a complex with a loosely bound partner that falls off in some particles. If you average all particles together, the partner region will be weak and ambiguous, tempting you to conclude the partner is disordered or absent. Classification can reveal two populations: one with clear density for the partner, one without. That difference is often biologically meaningful. It can reflect binding affinity, regulatory control, or the presence of competing factors in the preparation.

The emotional challenge is that classification is never perfectly decisive. Particles are noisy; assignments are probabilistic; boundaries between classes can be fuzzy. Some particles genuinely sit in between because the molecule passes through intermediate conformations. Others appear in between because the images are too weak to tell. This is where the metaphor of "piles" becomes slightly misleading. The best modern methods treat class membership as a likelihood rather than a verdict, and they may allow particles to contribute to multiple classes with different weights.

In recent years, the very idea of discrete classes has been questioned for some kinds of heterogeneity. Not all molecular motion is a flip between two states. Some motion is continuous: a domain swivelling through a range of angles, a loop breathing, a hinge sampling many positions. For this, forcing the data into a small number of classes can feel artificial, like describing a rainbow using only a few paint chips.

Neural approaches such as cryoDRGN, introduced by Zhong and colleagues (2020), aim to model this continuous variability more directly. Rather than producing a handful of separate maps, they can learn a latent "conformational space" in which each particle is a point, and nearby points correspond to similar structures. The promise is alluring: instead of choosing between open and closed, you might recover a continuum of opening. The cost is interpretability and, especially, validation. A method powerful enough to model a landscape is also powerful enough to create a plausible landscape that flatters the noise. The more freedom you give the model, the more disciplined you must be about asking what is truly supported by independent evidence.

Even within more traditional pipelines, the interplay between classification and refinement has become a craft. Researchers may perform 2D classification to clean the stack, then an initial 3D model, then a round of 3D classification to separate major states, then refined reconstructions for each state, sometimes followed by more focused classification on regions of interest. Masks can be used to concentrate on a flexible domain while ignoring the rest, a tactic that can reveal subtle motions without being drowned by the stable core. Each decision is an attempt to match the algorithm's assumptions to the biology's behaviour.

The craft is also a kind of humility. The earliest impulse in any imaging method is to demand a single definitive picture. Cryo-EM has had to learn, sometimes reluctantly, that the definitive answer may be plural. A "better" map is not always a single, sharper consensus; it may be a set of maps that, together, admit that the molecule is not locked in one pose. The crowd, properly sorted, can tell a richer truth than the crowd forced to speak in unison.

There is a final, practical payoff to this sorting mentality. Once you have separated particles into cleaner, more homogeneous groups, each group can be refined to higher quality. Alignment becomes easier because the reference is no longer a compromise. Averaging becomes more effective because the signal is more consistent. The result can be counterintuitive: by throwing away data or by dividing it into smaller subsets, you can obtain *more* detail. It is the statistical version of focusing a conversation. A room full of people talking at once contains more words, but less meaning. Separate them into discussions where participants share a topic, and the content becomes intelligible.

Cryo-EM began by asking a crowd to reveal what no individual could. As the method matured, it learned a second lesson: crowds are only wise when they are allowed to disagree.

4.4 How to Doubt a Beautiful Map

A cryo-EM map can be gorgeous in a way that makes the scientific part of the brain briefly go quiet. A surface rendered in soft greys, ridges and grooves catching virtual light, a suggestion of helices coiling with the neatness of a textbook drawing: it is easy to feel that the problem has been solved. The danger is that modern reconstruction pipelines are not only powerful; they are persuasive. They are designed to extract weak signal from heavy noise, and the same machinery that can reveal truth can also, if mishandled, polish coincidence into something that *looks* like truth.

Scepticism in this setting is not a personality trait. It is a technical requirement, because refinement is a feedback process. Once a reconstruction exists, it becomes the reference used to align particles more precisely; those better-aligned particles then produce a better reconstruction; and so on. This virtuous circle is the heart of the method. It is also the path to the central failure mode: *overfitting*, where the algorithm begins to align and reinforce noise as if it were signal. Overfitting does not announce itself by making a map uglier. It can do the opposite. A map can grow crisp, confident, and wrong.

It helps to be concrete about how the illusion can arise. A raw particle image is mostly noise, with a faint imprint of the molecule. When you align such images to a reference, you are always at risk of "finding" the reference in the noise. If you show someone a cloud and ask them to rotate it until it resembles a rabbit, they will eventually see a rabbit. A computer can do something similar, except it will do it for hundreds of thousands of clouds with relentless consistency. If the reference contains a small feature that is not truly in the data, alignment can still nudge particles into positions where random fluctuations resemble that feature. Averaging then reinforces the resemblance. After enough rounds, the feature can appear stable. The map begins to contain details that are reproducible only because the procedure has made them reproducible.

The community's most widely adopted defence is disarmingly simple: split the data in two. Instead of refining one map from all particles, you randomly divide the particle stack into two halves and treat them as independent datasets. You refine two maps, called *half-maps*, keeping the halves separate throughout refinement. This "gold-standard" approach was articulated clearly by Scheres and Chen (2012) as a safeguard against overfitting and shared-reference bias. The core intuition is everyday: if a detail is real, it should be supported by two independent sets of evidence. If it appears only when both halves are pulled toward the same reference, it is suspect.

Once you have two half-maps, you can compare them. The standard comparison is the *Fourier shell correlation* (FSC). The name sounds forbidding, but the concept is friendly. Any 3D map can be described in terms of spatial frequencies: broad, low-frequency components define the overall shape, while higher frequencies capture finer features. FSC asks, frequency by frequency, how similar the two half-maps are. At low frequencies the correlation is usually high because overall shape is robust; at higher frequencies the correlation falls because noise dominates and genuine fine detail becomes harder to reproduce. The curve is a way of locating the boundary where agreement ends.

This boundary is often summarised as "the resolution," quoted in angstroms. The number is useful, but it is also a magnet for misunderstanding. First, FSC is not measuring resolution in the way people learned it in school, as if the map were a photograph with a single focus setting. It is measuring reproducibility between two reconstructions under a particular processing pipeline. Second, the boundary depends on a criterion: at what correlation value do you declare that the map has stopped agreeing? The field has debated these thresholds for years. van Heel and Schatz (2005) discussed the logic and pitfalls of FSC criteria, and the existence of multiple commonly used cutoffs (0.5, 0.143, and others) reflects the fact that no single threshold is a law of nature. A fixed number can be misleading if the underlying assumptions are not met, or if processing steps have introduced correlations that masquerade as signal.

The most common way correlations are introduced artificially is by reusing information between the halves. This can happen accidentally, for example if a single reference leaks into both refinements in a way that couples them. It can also happen more subtly through masking and post-processing. A *mask* is a way to focus analysis on the region where the particle is, excluding empty solvent. Masks are essential for many steps, because solvent regions are noisy and can dilute statistics. But a mask can also produce spurious correlations by imposing a shared boundary. If both half-maps are multiplied by the same sharp-edged mask, they can become similar at high frequencies for purely mathematical reasons, even if the underlying densities are not truly consistent. The FSC can then look better than it deserves.

This is why cryo-EM validation developed a culture of cross-checks rather than a single talismanic plot. Chen and colleagues (2013) introduced *high-resolution noise substitution* as a practical diagnostic for overfitting. The idea is to deliberately replace the highest-frequency components of the particle images with noise and then run the reconstruction and refinement procedure. If the pipeline "recovers" high-frequency agreement between half-maps in frequency bands where the input was explicitly turned into noise, that is a red flag. It is a stress test: a way of asking whether the procedure can hallucinate detail when no detail exists to be recovered. Such tests are not about catching fraud; they are about catching the ordinary self-deception that powerful optimisation can create.

The need for these safeguards grew as cryo-EM entered the regime where maps began to hint at atomic structure. At moderate resolution, a wrong detail is often obviously wrong: a lumpy region where you expected a helix, or a smeared core where symmetry should be crisp. At near-atomic resolution, wrong details can be dangerously plausible. A groove can resemble a backbone trace. A spur can resemble a side chain. Map sharpening, a common post-processing step that boosts high-frequency contrast to make features more visible, can intensify this plausibility. Used well, sharpening reveals genuine structure that is present but muted; used aggressively, it can make noise sparkle like texture. The resulting surface can be aesthetically satisfying in exactly the way that should make a scientist uneasy.

This brings us to a second crucial point: resolution is not uniform across a map. Cryo-EM particles are not perfect rigid objects. Flexible regions move; peripheral domains may be less well aligned than a stable core; some orientations may be underrepresented, leaving certain directions less constrained. Even if the overall FSC suggests a particular global resolution, parts of the map may be better and parts worse. A single number can conceal this unevenness.

Local resolution estimation addresses that by asking, region by region, how far trustworthy detail extends. The result is often intuitive: the core of a stable complex may support crisp features, while a mobile arm fades into fog. Local resolution is more than a comfort; it influences what claims can be made. A binding pocket might be interpretable in the core but ambiguous in a flexible domain. A helix might be unbroken in one area and only a suggestion in another. Presenting a single global resolution without conveying this variation can invite readers to over-interpret the weakest parts of the map.

Sampling adds another constraint that can quietly set an upper bound. A reconstruction is computed on a grid of voxels, and the voxel size depends on the microscope magnification and detector pixel size. The *Nyquist limit* states, in effect, that you cannot reliably reconstruct details finer than twice the pixel size, because the data are not sampled densely enough. This is not a matter of algorithmic cleverness; it is mathematics. If you record data on a coarse grid, no amount of refinement can conjure frequencies you never sampled. Quoting a resolution better than the Nyquist limit is therefore a warning sign that something has gone wrong in calibration or reporting.

Orientation bias is another way a map can be beautiful and misleading. If particles prefer certain orientations on the grid, then projections from some angles are abundant while others are scarce. The reconstruction then becomes anisotropic: detail is stronger in some directions than others. The map may look sharp when viewed in one orientation and soft in another, yet a single FSC curve may not fully communicate this directional imbalance. Practitioners often diagnose the problem by inspecting view distributions and by using anisotropy-aware metrics, but for a general reader the key idea is simple: a sculpture built from photographs taken mostly from the front will have a well-defined face and an uncertain profile.

A practical vignette captures how these concepts collide. Imagine a team working on a membrane protein complex, desperate to understand how a small drug molecule binds. The refined map looks encouraging. After sharpening, a pocket appears to contain a neat, curved density that could be a ligand. The surface rendering is compelling; the story almost writes itself. But the half-map comparison tells a less romantic tale. When the two half-maps are examined separately, the "ligand" feature is stronger in one half and weaker in the other. The FSC at the highest frequencies is sensitive to masking choices. A noise-substitution test suggests that the pipeline can produce spuriously high correlations in the band where the ligand would live. Local resolution estimates show that the pocket sits in a region that is less stable than the core. Suddenly the beautiful density reads differently: not as a solved binding mode, but as a hypothesis. The team may still fit a ligand model, but it is now accompanied by caution, alternative interpretations, and perhaps an experimental plan to confirm binding by mutagenesis or biochemistry. Validation has not killed the story; it has made it honest.

These validation practices did not emerge by accident. As cryo-EM grew more central to structural biology, the stakes rose. A wrong model can waste years of downstream work or misdirect drug development. The community's response included explicit organising around standards and expectations. The EM Validation Task Force, discussed by Henderson and colleagues (2012), helped crystallise the idea that depositing maps and models should come with evidence about how trustworthy they are: reporting half-map procedures, describing processing choices, and using multiple complementary checks rather than relying on a single headline number. Later analyses and guidance, such as Lawson and colleagues (2021), continued this push toward transparency and annotation.

None of this makes cryo-EM uniquely fragile. Every method that extracts structure from limited data has its own illusions. X-ray crystallography has model bias and refinement traps; NMR has ambiguities and ensemble interpretations; even light microscopy can over-interpret deconvolved images. What is distinctive about modern cryo-EM is that its images are so noisy at the particle level and its computational machinery so capable that the method can produce convincing artefacts unless the workflow is designed to prevent them.

There is a humane way to think about what validation really does. It does not ask whether a map is perfect. It asks whether the map's details are *earned*. Earned details survive when the data are split, when processing choices are varied within reason, when masking is treated with care, and when claims are limited to regions where the map genuinely supports them. Earned details also align with the physics of sampling: they do not exceed the Nyquist limit, and they do not magically appear in directions where the data provide few views. An earned map can still be incomplete; it can still have flexible regions; it can still resist interpretation in places. But it does not pretend.

In the end, the discipline of doubt is what allows cryo-EM to be ambitious without becoming reckless. The most trustworthy maps are not those that look the smoothest or the sharpest in a rendering, but those that remain themselves when asked, repeatedly and from different angles, to justify their beauty.

Chapter 5

Putting Atoms Where the Fog Used to Be

The miracle of cryo-EM is not simply that it makes the invisible visible, but that it turns shadows into arguments about how life works. In these pages, the reader watches scientists decide where an atom belongs, where confidence ends, and how a good structure becomes the start of the next experiment rather than the last word.

5.1 Density Is Not a Molecule

The first time most people see a cryo-EM reconstruction, the temptation is immediate: treat it as an object. A grey, lumpy volume floats on the screen, and with the right contour it even has the satisfying solidity of a sculpture. You can rotate it, slice it, zoom into a groove that seems made for a small molecule to nestle inside. It feels natural to say, *there's the protein.* Yet that sentence sneaks in an assumption that cryo-electron microscopy constantly tries to punish: a map is not the molecule. It is a calculated estimate of where signal accumulates in three-dimensional space after many noisy glimpses have been combined. It can be wonderfully revealing, and still be uncertain in exactly the places a human mind most wants certainty.

A cryo-EM map begins far upstream of anything that resembles a tidy shape. In single-particle cryo-EM, the microscope does not image a single molecule continuously from many angles, the way a medical CT scanner circles a patient. Instead it records a field of view sprinkled with thousands of copies of the same molecule, each frozen in a thin film of glassy ice and each lying at its own unknown orientation. The raw images are not portraits. They are faint, interference-rich projections in which the molecule's contribution is often barely more obvious than a smudge in a grainy night photograph. The crucial trick is that all those smudges contain overlapping information: different views of the same underlying structure. Computation supplies the missing geometry, aligning and averaging until what is shared rises above what is random.

Even that seemingly straightforward phrase, *aligning and averaging*, hides the most important idea a map has to teach: the map is a statistical compromise. Every particle image is a bargain between signal and noise; every alignment is a best guess; every average is an attempt to keep what is consistent while letting the rest cancel out. If you could replay the entire reconstruction with a different random subset of particles, you would not get an identical map. The differences might be tiny at high confidence, or substantial in flexible regions, but they are always there. A map is therefore less like a photograph and more like a weather forecast: a carefully constrained prediction based on incomplete, noisy observations, produced by a method that is itself transparent enough to scrutinize and yet complex enough to surprise.

There is another subtlety that matters once one begins to think about placing atoms. X-ray crystallography traditionally speaks of *electron density*, a map of where electrons are likely to be found. Cryo-EM single-particle maps are usually described instead as *electrostatic potential*. The distinction can sound pedantic, but it shapes expectations. In very simple terms, the contrast in an electron microscope depends on how the specimen's charges and atoms affect the phase of the electron wave. What emerges from the reconstruction is a three-dimensional field that correlates with the molecule's electrostatic landscape, not a direct count of electrons sitting in particular orbitals. For many practical purposes, proteins in cryo-EM maps still display familiar features—backbones, side chains, helices, and sheets—but the map is not a literal inventory of matter. It is a physically grounded signal transformed by the instrument, by the specimen, and by computation.

The way maps are displayed makes the illusion of solidity stronger. A typical cryo-EM map is shown as an *isosurface*: a skin drawn through the 3D field wherever the value crosses a chosen threshold. Turn the threshold up, and only the strongest regions remain: dense cores, sturdy helices, rigid domains. Turn it down, and the object swells, sprouting bridges and tendrils that may represent real features, noise, or the statistical echo of misaligned particles. Choosing the contour level is not cosmetic; it is an interpretive act. At a dinner party, one might say: it's like sculpting a coastline from satellite data. With a strict cutoff, the land is tidy but many shallow bays disappear. With a generous cutoff, you keep every inlet and sandbar, but you also mistake fog for shore.

Working scientists learn to read a map the way an experienced sailor reads the sea: not by trusting a single view, but by looking for consistency across contexts. A helical bundle that remains crisp across different contours is rarely an illusion. A wispy tube that appears only at one threshold might still be real—a flexible tail that only occasionally occupies a position—or it might be the map politely offering you a mistake. The human brain is exceptionally good at seeing patterns, including patterns that are not there. Cryo-EM is an arena where this talent is both essential and dangerous.

A concrete example brings the point home. Consider a membrane receptor with a floppy intracellular loop. In a living cell, that loop may swing widely, contacting different partners at different times. When thousands of copies are frozen for cryo-EM, each loop is captured in one moment, but not the same moment. The structured core of the receptor aligns well: the transmembrane helices are like a rigid handle. The loop, by contrast, refuses to sit still across the dataset. During averaging it becomes diluted, smeared, or disappears into a vague haze. The resulting map is not wrong; it is honest about what the data support. Yet it is also incomplete in a very specific way: it does not say the loop is absent from the molecule. It says the loop does not have a single, well-defined position shared across the particles used to compute this particular reconstruction.

That honesty can be hard to keep in mind once one starts turning a density map into an atomic model, because the mind dislikes blankness. A map with a crisp core and a foggy periphery invites a kind of narrative filling-in: perhaps the loop runs *here*, perhaps it packs against *that* helix, perhaps it forms a short α-helix before vanishing. Sometimes biochemistry or prior structures justify such guesses. Sometimes the guesses harden into unearned certainty, especially when an atomic model is drawn as a continuous chain that cannot simply evaporate into statistical uncertainty. The chain, once drawn, carries rhetorical force: readers see it as a claim about reality rather than as a hypothesis perched on weak support.

Cryo-EM developed a culture of caution partly because of its own history. In the late twentieth century, three-dimensional reconstructions from electron microscopy were sometimes dismissed as artistry: impressive images, but hard to trust at atomic precision. One turning point came from the logic, as much as the technology, of single-particle analysis. Joachim Frank and colleagues helped establish, over decades, the mathematical framework that made it possible to reconstruct a 3D object from many noisy projections without crystallizing the specimen. The payoff was not merely prettier pictures, but an explicit way to attach uncertainty to them: if the same structure emerges from independent subsets of data, it is less likely to be a shared hallucination. Modern practice formalized this idea with routines that split the dataset into two halves, reconstruct two maps independently, and compare them across spatial frequencies—a way of checking how much of what you see is supported by repeatable signal rather than by overconfident fitting.

Even with that discipline, a single number like "3.2 Å resolution" can be misleading. Resolution in cryo-EM is rarely uniform. A ribosomal core might be sharp enough to show the zig-zag of the backbone and the orientations of side chains, while a peripheral stalk is a soft cloud. A viral spike protein might have a stable base and a head that breathes between conformations. If you average everything into one global claim, you risk encouraging the wrong kind of reading: as if every region is equally trustworthy. Practitioners talk about *local resolution* to capture the reality that some parts of the map are drawn from many consistent observations and others from fewer or more variable ones. In human terms: you may know your friend's face with certainty, but only have a vague impression of the pattern on their scarf because it kept shifting.

The way maps are processed can also move the boundary between revelation and uncertainty. A raw reconstruction often appears smoother than one would like, partly because high-frequency information is weak. To compensate, scientists use *sharpening*, a family of methods that boost high-resolution terms so that edges and small features stand out. Used well, sharpening is like adjusting contrast in a dim photograph so that the texture of fabric becomes visible. Used aggressively, it turns noise into apparent structure: the grain becomes pores, the speckle becomes side chains. The difficulty is that different parts of a map may *want* different treatment. A rigid core can tolerate stronger sharpening; a flexible region may disintegrate into salt-and-pepper artifacts if pushed the same way. Any single, global knob-setting is an imperfect compromise, and compromises have consequences.

This is where the phrase "disciplined imagination" earns its keep. Model builders have to imagine, because a map is not self-interpreting. But they must imagine in a way that can be audited. The practical craft is full of small, almost tactile decisions: whether a tube is a helix or two strands close together; whether a bulge belongs to a bulky side chain or to a glycan; whether a break in density indicates a real gap, a flexible hinge, radiation damage, or simply the map falling below threshold. Each decision can be guided by chemistry and by the protein's sequence. If the sequence says there must be a proline, you anticipate a kink. If it says glycosylation sites abound, you are alert for sugar-like density. Prior knowledge is not cheating; it is part of the inference. But prior knowledge can also become a ventriloquist. A scientist can end up seeing what the protein *ought* to do, not what the data can support.

A famous, everyday version of this cognitive trap is reading blurred text. Show someone a smudged word in a sentence, and they will often "see" the correct letters even if the ink is absent, because context supplies the answer. Cryo-EM map interpretation is like reading an entire novel under water. Context is essential, and context can lie. The discipline is to keep asking: is this letter actually on the page, or am I supplying it?

Contour levels provide a simple illustration of how quickly confidence can be misplaced. Suppose a pocket in the map seems to contain a small ligand. At a certain threshold, a blob sits neatly in the cavity. Turn the threshold slightly, and the blob merges with the protein surface, losing its separateness. Turn it again, and the blob fractures into two specks. Which version is "real"? The honest answer is that the map does not contain a single, crisp object labeled "ligand." It contains a distribution of values that you are sampling with a contour. Real ligands can behave exactly like this if their occupancy is partial or their position variable. Noise can also behave like this. Deciding requires more than a satisfying visual: it demands consistency with chemistry, with known binding modes, with the electrostatic environment, and with the broader pattern of the reconstruction.

Even the most seductive feature in high-resolution maps—the apparent visibility of side chains—comes with built-in caveats. Some amino acids have bulky shapes that produce recognizable density; others are more ambiguous. Aromatic rings can be distinctive; a small polar side chain can hide in the blur. Moreover, electrostatic potential does not emphasize all atoms equally. Atoms with different scattering properties and local environments may produce different contrast, and nearby features can merge. This is why experienced builders speak carefully: they may say the backbone is well supported but the side-chain *rotamer* (the chosen orientation among common possibilities) is uncertain; or that the sequence register along a helix is clear in one region but could slip by one residue elsewhere. Such statements can sound like hedging until one realizes they are the difference between a model that supports experiments and a model that quietly misleads them.

A historical anecdote from the "resolution revolution" captures how perception can outrun evidence. When direct electron detectors and improved algorithms began producing markedly sharper reconstructions in the early 2010s, the field experienced a rush of near-atomic maps. Structures that had been foggy for decades suddenly displayed crisp helices, separated β-strands, and, in good cases, discernible side chains. The excitement was real and justified: new biology became legible. Yet many groups also discovered, sometimes painfully, that a beautiful map can tempt over-interpretation. A reconstruction might show convincing texture in one area while another area, processed with the same sharpening, acquires plausible-looking bumps that are actually amplified noise. The lesson was not that the new tools were untrustworthy, but that they made it easier to create a believable fiction if one forgot that every map is an estimate with an error model.

The most reliable way to keep the distinction clear is to treat the map and the model as different kinds of objects with different obligations. The map's obligation is statistical: to represent, as faithfully as possible, the signal that is reproducible across the particle images given the assumptions of the reconstruction. The model's obligation is chemical and physical: bond lengths must make sense, atoms cannot overlap impossibly, peptide chains must form plausible conformations, charged residues must sit in believable environments. When these obligations agree, confidence grows. When they conflict, the conflict is informative. Perhaps the map is too weak locally and the model should retreat to a simpler representation. Perhaps the model is being forced by expectation and should be loosened. Perhaps the molecule truly adopts multiple states and no single model should be asked to represent them all.

The phrase "putting atoms into density" can therefore mislead. It implies that atoms are *in* the map the way raisins are in a bun, waiting to be picked out. In reality, atoms are a hypothesis laid onto a field of reconstructed values. At very high quality, the hypothesis can be so strongly supported that it feels inevitable. At intermediate quality, it is more like tracing a coastline from misty aerial photographs: you can draw it, but you should mark where the shoreline is certain and where it is marshland. The most honest models are those that do not pretend the entire molecule was equally visible. They have confident cores and cautious edges; they acknowledge mobility; they avoid ornate claims where the data offer only a suggestion.

That restraint is not a loss of ambition. It is what gives a structure its peculiar power: the ability to turn a blur into knowledge without turning uncertainty into mythology. A cryo-EM map is a conversation between reality and inference. Listening well means resisting the urge to declare every bump a fact, and learning instead to hear, in the texture of the density itself, where the molecule speaks clearly and where it only whispers.

5.2 From Fitting to Building

A cryo-EM map can feel like an invitation: here is the hollow, here is the ridge, here is a helix-shaped tube running exactly where a helix ought to run. If biology were always obliging, turning that invitation into an atomic model would be a simple act of matching shapes. Sometimes it nearly is. Many molecules belong to large families, and evolution is a relentless recycler. A kinase domain from one species resembles a kinase domain from another; an antibody fragment has a recognizable architecture; the barrel of a chaperonin announces itself with symmetry. In those cases, the first move is often to borrow: take a known structure, or a predicted one, and place it into the map.

That move has a name that captures its mechanical feel: *rigid-body fitting*. "Rigid-body" means the model is treated as a solid object, allowed to rotate and translate but not to bend. It is the structural equivalent of putting a key into a lock to see if the outline matches. When it works, it feels almost magical. The borrowed model slides into the density with an obvious rightness; helices line up with helical tubes, sheets fill sheet-like slabs, a domain that once floated abstractly in a database now occupies a specific place in a working machine.

Yet rigid-body fitting is only the beginning, and not only because proteins are rarely as rigid as the phrase suggests. The deeper reason is that a cryo-EM map is not asking for *a* shape; it is asking for the *correct* shape in the *correct* state. A domain may swivel relative to its neighbours. A loop may adopt a conformation that has never been seen before. A helix may unwind near a ligand. A borrowed model can also be subtly wrong in a way that is hard to see at first glance: a register shift, where the backbone trace is correct but the sequence is slid by one or more residues along a helix or strand. The result is a model that fits the general density while putting the wrong side chains into critical positions—a mistake that can derail functional conclusions without ever producing an obviously "bad-looking" structure.

The craft begins at the moment that fitting stops being enough and building begins. Building is slower, more argumentative, and more intimate. Instead of moving a whole domain, the modeller negotiates residue by residue. Where exactly does the backbone run? Which way does a peptide bond point? Does this bulge correspond to a bulky aromatic side chain or to a turn in the chain? Can the sequence be read from the density with any confidence, or is the map too ambiguous there? These questions do not have the clean finality of a jigsaw puzzle. They are closer to translating an ancient text from a damaged manuscript: one must work with partial letters, repeated motifs, and the logic of grammar, always aware that confident guesses can be wrong.

A concrete scenario shows how quickly the work shifts from geometry to judgement. Imagine a medium-resolution map of a human membrane protein complex, perhaps around the range where helices and strands are recognizable but side chains are uneven. A predicted model from AlphaFold, or a homologous structure from a related organism, gives a strong starting point. The transmembrane helices dock well into the map: a reassuring sign that the overall fold is right. But as soon as you move to the extracellular loops, the agreement frays. A loop that should connect two helices appears to cut across a gap, while a nearby arc of density seems to be begging for a different path. If you simply force the predicted loop into place, you can obtain a continuous chain that "matches" the map in the sense that it sits inside the contour. The price is subtle strain: bond angles bend, side chains clash, and the loop's chemistry starts to look improbable. If you instead allow the loop to be rebuilt, a new path may emerge that respects both density and protein geometry. The map did not contain a loop as a ready-made object; it contained enough information to constrain plausible loops, and the modeller's job is to choose one that survives both the data and the laws of molecules.

This is where tools such as *Coot* earned their reputation. Coot is often described as model-building software, but that phrase understates the experience. It is closer to a workbench where you can grab a segment of chain, nudge it, twist it, flip a side chain into a different rotamer (a common, energetically preferred orientation), and instantly see whether the model becomes more chemically sane without losing contact with the map. That immediacy matters. It turns modelling into a conversation: density suggests a move; chemistry accepts or rejects it; the modeller tries again.

The conversation becomes more complicated once the map itself is treated, because map processing can either make the right answer visible or make the wrong answer persuasive. *Sharpening* is the classic example. Cryo-EM reconstructions often benefit from boosting high-resolution information so that boundaries and small features stand out. The danger is that sharpening is not a free lunch: it amplifies noise as well as signal. The human eye is exquisitely sensitive to crisp edges, so a sharpened map can feel more authoritative even when it is simply more textured. A modeller may "see" a side chain peak because the map has been sharpened into a spiky landscape, then place the side chain accordingly, then feel reassured because the side chain now sits on a peak that was partly created by the processing choice. The result is not fraud; it is a feedback loop between expectation and display.

Experienced groups often respond by treating the map as plural rather than singular. They inspect alternative reconstructions, different sharpening levels, and sometimes different ways of weighting the data, asking whether a feature survives changes that should not affect genuine signal. They may keep a conservative map for building the overall chain and a more aggressively processed one only for checking specific details. In practice, this means cultivating a kind of visual humility: a peak that appears only under one set of settings should not carry the same confidence as a feature that keeps returning no matter how the map is presented.

When no adequate starting structure exists, the work becomes *de novo* model building: tracing the chain largely from scratch. At its best, this feels like solving a mystery with multiple kinds of clues. The map provides shape clues: helices show up as rods; β-sheets as flattened layers; the spacing between strands can hint at direction. The sequence provides pattern clues: glycine and proline often mark turns; long hydrophobic stretches suggest membrane segments; a run of bulky residues might produce a distinctive "lumpiness" if the map is good enough. Biochemistry provides plausibility clues: disulfide bonds are constrained; active sites tend to conserve motifs; charged residues are unlikely to be buried without compensation.

The most treacherous trap in de novo building is the seduction of continuity. A protein chain in a textbook is a continuous ribbon. A real cryo-EM map may contain interruptions: density that fades in flexible regions, breaks where multiple conformations have been averaged, or gaps where local alignment was poor. Faced with a break, the modeller has options: leave the region unmodelled; represent it as a simplified poly-alanine backbone (a way of acknowledging a traced path without asserting side-chain identities); or attempt to bridge it with a specific loop and sequence assignment. The first option feels unsatisfying but is often the most honest. The second is a compromise. The third is tempting, and sometimes correct, but is also where quiet distortions begin: a wrong bridge can drag the chain register out of alignment for dozens of residues beyond the gap.

The history of structural biology contains instructive examples of how such distortions echo. Early protein crystallography, long before modern validation became routine, produced occasional structures with improbable geometries that nevertheless seemed to "fit" the electron density of the time. Those models were not the product of incompetence; they were the product of sparse data, enthusiastic inference, and the understandable desire to complete the story. Cryo-EM, despite its different physics, faces a related pressure. A partial model can feel like an unfinished sentence. But an over-complete model can be worse, because it reads as confident prose when it is actually conjecture.

One way the field has tried to lower the temperature of this guessing game is by making the model obey more than the map. Traditional fitting can, in the worst case, treat the model as soft clay that must be pressed into density at any cost. Physics-based and sampling-based approaches push back. *RosettaES*, for example, uses energetic realism and extensive sampling to explore alternative conformations that could plausibly occupy the density, rather than accepting the first visually convenient trace. The point is not to let an energy function overrule the data, but to prevent models that satisfy the map while violating nonbonded interactions, packing, or backbone preferences. In the ambiguous resolution range where the backbone may be clear but peptide orientations and side-chain placements are not, that kind of energetic discipline can be the difference between a model that is merely plausible and a model that is chemically coherent.

A particularly vivid modern tool for this is *ISOLDE*, which brings interactive molecular dynamics into the model-building process. Molecular dynamics is a simulation technique that treats atoms as objects subject to forces, allowing them to move in physically motivated ways. In an interactive setting, the modeller can tug on a loop or rotate a side chain, and the rest of the structure responds as a real molecule would: bonds resist stretching, clashes create repulsion, hydrogen bonds form or break. The experience is revealing even for non-specialists. It exposes how easy it is to create a visually tidy model by forcing atoms through each other, and how quickly physical realism pushes back. A structure that "fits" the map but cannot exist without severe strain is a warning sign, not a triumph.

Automation has also changed what "building" means. Packages such as *Phenix map_to_model* can take a cryo-EM map (within practical limits of resolution and quality) and produce a substantial atomic model with far less manual tracing than was once required. When these pipelines work well, they are transformative: they free experts to spend time on the genuinely hard interpretive decisions rather than on routine backbone tracing. But they also shift the location of judgement rather than eliminating it. Automated building depends on preprocessing steps such as sharpening and segmentation (deciding which part of the map corresponds to a particular molecule or domain). It relies on libraries of known structural patterns and on assumptions about what protein geometry usually looks like. If the map contains unusual chemistry, rare conformations, or mixed states, the automation can confidently produce a model that is *typical* rather than *true*.

Machine learning has pushed that boundary further. *ModelAngelo* and related approaches integrate map signal with learned statistical expectations from known structures and with sequence information. The promise is striking: faster building, better handling of noisy regions, even the ability to identify unknown proteins in a complex mixture by matching sequence patterns to density. The risk, again, is subtle: when a tool uses strong priors, it can fill in uncertainty with what is most probable *in general*. That is powerful when the molecule behaves like most molecules. It is hazardous when biology does something rare, or when the experimental map is weak in exactly the region that distinguishes one hypothesis from another.

A memorable real-world example of why this care matters arrived in public view during the rush to understand SARS-CoV-2. Cryo-EM structures of the coronavirus spike protein quickly became central objects in global science: they guided vaccine design, antibody discovery, and a flood of mechanistic speculation. The spike is also a protein that likes to move. Its receptor-binding domains can hinge up and down, and parts of it are decorated with flexible glycans. In practice, this meant that some regions of the spike could be built with great confidence while others could not. The most responsible models were the ones that resisted the urge to "finish" every glycan, every loop, every edge of a moving domain. In a context where millions of people were waiting for answers, the discipline of leaving uncertainty visible became an ethical choice as well as a technical one.

At the bench, the modeller's work often ends up as a patchwork of confidence. A stable enzyme core may be built with side chains placed and refined until they sit naturally in peaks of density. A peripheral helix may be traced but left with simplified side chains. A flexible tail may be omitted entirely, not because it is unimportant, but because the map cannot support a single, specific placement. This patchwork is not a flaw; it is a truthful reflection of what the reconstruction can and cannot say.

The craft, at its best, is not about forcing a molecule to appear, but about earning the right to make atomic claims. A well-built model does not merely occupy the density; it explains it without bullying it, and it remains chemically believable even when no one is watching. The reward is a structure that can be used as a working object in biology: a binding pocket whose residues can be mutated with purpose, an interface whose contacts can be tested, a hinge whose motion can be probed. That usefulness depends on a quiet kind of restraint: the willingness to stop building exactly where evidence stops, and to let the remaining fog stay foggy rather than turning it into stone.

5.3 Validation as a Scientific Immune System

An atomic model can be dangerously convincing. Once atoms are drawn, coloured, and rendered with smooth shading, the mind stops seeing a hypothesis and starts seeing a thing. That is not a flaw in the viewer; it is a feature of human perception. We are pattern-making animals, and a continuous chain with tidy helices and neatly packed side chains triggers the same sense of completion as a finished sentence. In structural biology, that sense of completion is precisely what must be challenged.

Validation is the set of habits and tests that provide the challenge. It is not there to spoil anyone's enjoyment of a beautiful structure, or to enforce a bureaucratic standard. It functions more like an immune system: it patrols for attractive mistakes, the kind that spread easily because they look plausible. When validation works well, it catches errors early, when they are still local and correctable. When it fails, errors can circulate for years, shaping experiments, drug campaigns, and entire stories about how a molecule works.

The need for validation is built into the way cryo-EM models are made. As we saw, the map is a statistical compromise, and model building adds further choices: sharpening, sequence assignment, whether to build a loop or leave a gap, whether a blob is a ligand or merely a fluctuation. A modeller can always improve the apparent fit to a map by making the model more contorted. If all that mattered were agreement with density, the winning strategy would be to squish the chemistry until it hugs every bump. The basic premise of validation is that good models must satisfy *two* kinds of truth at once: truth to the experimental reconstruction and truth to the known constraints of molecular structure.

Start with the most intuitive family of checks: *map–model agreement*. If the model is meant to explain the map, it should correlate with it. Software reports various correlation coefficients (often described broadly as CC metrics) that quantify how well the model's calculated signal matches the experimental map. These numbers are useful, but they are also easy to misread. A single global correlation can be high even when particular regions are unsupported, because strong, well-resolved parts dominate the statistic. It is the same problem as averaging grades across a school: excellence in one class can hide failure in another.

More subtle is the danger of circularity. If a map has been improved using assumptions that resemble the model, or if a model has been refined so aggressively that it begins to follow noise, map–model agreement can become a measure of self-consistency rather than correctness. Cryo-EM developed a powerful antidote to this kind of self-reassurance: *gold-standard half-map procedures*. The underlying idea is almost childishly simple. Split the data into two independent halves. Reconstruct two maps separately. Only trust features that are reproducible between them. This independence principle, used widely in modern workflows, makes it harder for spurious correlations to masquerade as signal. It also provides a way to estimate resolution without letting the reconstruction chase its own tail.

Once atoms enter the picture, an analogous kind of independence becomes desirable: the model should not only match a map, it should match the *right* map in the *right* way, without borrowing confidence from the very choices it is supposed to justify. One common approach is to compare the model to maps derived from different halves of the data, or to compute a map–model Fourier shell correlation (map–model FSC) that reports how well the model is supported across spatial frequencies. These tools can reveal an uncomfortable truth: a model can track low-resolution features beautifully while being essentially unconstrained at higher resolution, where individual side-chain choices live.

Map agreement, however, is only half of the immune system. The other half is *geometry and chemistry*, the internal health checks that ask whether the model makes sense as a molecule. Here, structural biology benefits from a long history, especially from X-ray crystallography, where the community learned that seductive models can be geometrically ill. Jane S. Richardson and David C. Richardson's work on MolProbity, continually expanded and formalized (with a major reference update in 2018), turned "chemical common sense" into quantitative tests. Are bond lengths and angles reasonable?

Do side chains adopt plausible rotamers, or do they sit in strained, rarely observed orientations? Are there steric clashes, atoms unrealistically colliding as if matter were transparent? Does the backbone occupy allowed regions of the Ramachandran plot, the map of sterically feasible backbone angles?

These checks have a particular flavour: they are not derived from the cryo-EM experiment at all. They come from accumulated knowledge of how proteins behave. That independence is precisely why they are useful. A model can always be bent to satisfy density; it cannot easily be bent to satisfy density *and* avoid clashes *and* maintain realistic backbone geometry unless it is close to correct.

A vivid vignette, familiar to many modellers, shows how this plays out. Suppose a researcher has a map at intermediate resolution and is eager to place an active-site loop that appears as a broken ribbon of density. With some careful pushing in an interactive program, the loop can be made to connect two well-defined anchors. The chain becomes continuous, the model "looks right," and the map correlation rises. Then MolProbity reports an explosion of clashes in the loop and a cluster of Ramachandran outliers: backbone angles in forbidden territory, like a gymnast's limbs rotated beyond human joints. The instinctive reaction is to suspect the validation tool. But more often the tool is pointing at a genuine problem: the loop has been forced through space in a way the protein cannot adopt. The model is not a discovery; it is a confession of overenthusiasm.

Cryo-EM adds a twist, because many reconstructions fall in a resolution range where the backbone trace is visible but the orientation of peptide bonds is ambiguous. The peptide plane contains the carbonyl oxygen, and its direction affects hydrogen bonding, secondary structure, and function. At around 2.5 to 4 Å, a modeller may place the backbone correctly in a broad sense while flipping peptide orientations in ways that the map cannot clearly contradict. This is exactly where CaBLAM becomes valuable. CaBLAM is a validation method designed to flag backbone pathologies in cases where the peptide orientation is hard to read directly. It does not require the map to show every atom; it checks whether the implied backbone geometry behaves like real protein backbones do. In immune-system terms, it is a specialist cell type: not always needed when everything is crystal clear, but crucial in the grey zone where plausible mistakes are common.

Feature-level validation goes even further, asking whether specific kinds of detail are supported rather than trusting global scores. A well-known example is EMRinger (2015), which focuses on side chains. Side-chain placement is one of the most satisfying moments in building: a rotamer clicks into place, and the model suddenly seems alive. It is also a moment of vulnerability, because side chains can be "explained" into noise. EMRinger provides an orthogonal check by assessing whether the map has the expected signal where side-chain conformations predict it should be. If many side chains land in orientations that are systematically unsupported, that is a warning that the model is being guided more by expectation than by data, or that the map processing has created misleading texture.

Then there is the question that haunts almost every headline resolution claim: *where* is the structure reliable? A single number for resolution, or a single map–model correlation, can conceal the truth that some regions are solid while others are little more than conjecture. This is why local metrics have become so important. Q-score, introduced by Grigore Pintilie and Wah Chiu (2020), aims to quantify atom resolvability: a way to ask, at the level of atoms or residues, whether the surrounding map supports the presence and position of that atom. FSC-Q (2021) addresses a related need by focusing on local support for the model based on local Fourier shell correlation concepts. Both tools respond to the same practical reality: it is possible to build a model that is exquisitely accurate in one domain and largely speculative in a flexible tail, and readers deserve to know the difference.

This local perspective changes how one reads a structure. Imagine a receptor bound to a small drug-like molecule. The core of the receptor might have strong local scores and clean geometry. The pocket might be convincing. But the ligand itself could be in a region of weaker density, perhaps because it binds with partial occupancy or adopts multiple poses. Global metrics may still applaud, because the protein dominates the calculation. A local check can be more candid: it might say, in effect, "the protein is well supported here, but the ligand placement is not." That difference is not academic. If a medicinal chemist uses the ligand pose to design new compounds, overconfidence in the ligand placement can waste months of work.

Validation also intersects with social reality: structures are not private objects; they are public claims. Cryo-EM models are deposited in archives, re-used by other groups, and incorporated into textbooks and drug discovery pipelines. That means the immune system must include community scrutiny. Peer reviewers ask for maps, half-map comparisons, and validation reports. Databases increasingly compute standardized metrics to make entries comparable. Laboratories develop shared norms: leaving weak regions unmodelled rather than inventing detail, reporting local resolution variability, and distinguishing clearly between what is supported and what is inferred.

None of this eliminates judgement. A validation outlier is not automatically an error; biology sometimes uses strained conformations for function, and unusual backbone geometries can occur in tight turns or active sites. A perfect validation report does not guarantee truth, because a wrong model can be geometrically plausible. The point is that validation forces the modeller to *account* for claims. If a loop has outliers, there should be a reason. If a ligand is placed, there should be local support. If side chains are modelled, the map should carry the signal that justifies them. The immune system does not prove health; it makes disease harder to ignore.

There is a quiet emotional benefit to this discipline. Without validation, model building can become a contest between the scientist and the fog: a struggle to extract certainty from ambiguity. With validation, the struggle becomes more constructive. The question shifts from "can I make this fit?" to "can I make this fit *honestly*?" The reward is a structure that can be trusted not because it is pretty, but because it has survived multiple, independent attempts to find what is wrong with it—and has come through with its most vulnerable claims either defended by evidence or left, openly, as uncertain.

5.4 What a Structure Lets You Ask Next

A reliable atomic model changes the kinds of sentences biologists can write. Before, they speak in broad gestures: a protein "binds" a partner, a channel "opens," an enzyme is "regulated." After, the verbs acquire handles. Binding becomes a set of contacts between particular side chains. Opening becomes a hinge rotation around a defined pivot, a gate formed by specific residues, a constriction that widens by a measurable amount. Regulation becomes a network of salt bridges and hydrogen bonds that rearrange as a ligand arrives. The structure does not end the argument about how life works; it makes the argument precise enough to test.

The most direct new question is embarrassingly simple: *where does something bind?* Cryo-EM has made this question answerable for many large, dynamic targets that resisted crystallography, including membrane proteins and sprawling molecular machines. In the atomic model, a binding pocket is no longer a vague cavity. It has an entrance, a floor, side walls, an electrostatic personality. A chemist can see whether the pocket is mostly hydrophobic (greasy, favouring nonpolar ligands) or lined with charged residues (favouring polar groups, salt bridges, and water-mediated contacts). A biologist can see whether the pocket is deep and selective or shallow and promiscuous.

With that visibility comes the ability to design mutations that do not merely "perturb the protein," but probe a specific idea. If a ligand appears to be anchored by a particular arginine, mutate it to alanine and ask whether binding drops. If a pocket has two plausible poses for a drug-like molecule, choose mutations predicted to favour one pose over the other. The structure becomes a map for experiments, but only if the confidence in those atomic placements is treated with respect. Granular confidence metrics, such as per-atom or residue-level Q-scores and local fit measures, can act like a practical conscience: they highlight which regions of the model are reliable enough to support precise claims and which are still scaffolds built into uncertain density. A mutation planned around a high-confidence side-chain network is a sharper test than one aimed at a loop whose local metrics say, politely, "don't bet your grant on this."

A second class of questions emerges when structures are captured in more than one state. Even a pair of states—ligand-free and ligand-bound, closed and open, inactive and active—is enough to turn the model into a mechanical hypothesis. Where is the hinge? Which helix tilts? Which interface breaks? Sometimes the motion is dramatic, like a jaw opening. Often it is subtle: a half-turn of a side chain that triggers a cascade of rearrangements, like one misbuttoned shirt pulling the whole front askew. The value of atomic resolution is that it lets you trace these cascades as physical chains of cause and effect rather than as metaphors.

Consider a generic but common motif: allostery, regulation at a distance. A ligand binds in one pocket, yet activity changes at another site tens of angstroms away. Without structure, allostery can sound mystical. With structure, it can become a set of linked constraints: a ligand pulls two helices closer, which tightens a salt bridge, which nudges a loop, which repositions a catalytic residue. The hypothesis is no longer "the protein senses binding." It is "this network transmits binding." That difference is profound, because networks can be tested node by node. If a salt bridge is central to the proposed pathway, mutating one partner should weaken the coupling. If a hinge is critical, stabilizing it with an engineered disulfide bond should reduce motion. A structure turns an abstract story into a set of specific predictions that can be falsified.

A third question is social rather than solitary: *how do molecules recognize each other?* Many of the most consequential events in cells are meetings: an antibody clasping a virus, a receptor engaging a hormone, a ribosome grabbing a transfer RNA, a chaperone capturing an unfolded client. Cryo-EM excels at these encounters because it can image large complexes without forcing them into crystals. The atomic model reveals not only that two partners touch, but how. Are there complementary charges like tiny magnets? Is the interface a smooth, well-matched surface or a few "hot spot" residues doing most of the work? Is water excluded, creating a dry, tight seal, or does a hydrated interface allow sliding and rearrangement?

Once recognition is understood at this level, intervention becomes more purposeful. Antibody engineering, for example, is not only about making binding stronger; it is also about making binding smarter. If an antibody neutralizes a virus by locking a protein in a particular conformation, then residues that stabilize that lock become obvious targets for affinity maturation. If a viral mutation breaks an antibody's grip by removing one key contact, the structure can suggest compensatory mutations in the antibody to restore binding. This kind of reasoning became a public-facing reality during the COVID-19 pandemic, when cryo-EM structures of the SARS-CoV-2 spike protein and its complexes with antibodies and receptors helped explain why certain variants evaded immunity and how antibody cocktails could be designed to reduce escape. The structural insight was not merely descriptive; it guided decisions about which antibodies were likely to remain effective and which epitopes were most vulnerable to mutation.

There is also a quieter, older historical arc behind these modern stories: the rise of structural thinking as a way to *ask* biological questions. When early protein structures began to appear in the mid-twentieth century, they shifted biology from a discipline of parts lists toward a discipline of mechanisms. An enzyme was no longer just something that accelerated a reaction; it was a shaped object with a site that could position substrates, stabilize transition states, and choreograph proton transfers. That transformation is happening again and again as cryo-EM brings previously inaccessible targets into focus. A membrane transporter is no longer a black box that "moves molecules across membranes"; it becomes a cycle of conformations with gates, cavities, and couplings that can be probed.

In practice, the most productive structures behave like experimental planning tools. A lab might begin with a rapid automated build—perhaps aided by modern ML approaches such as ModelAngelo—then immediately perform validation triage: geometry checks, map—model agreement, and local confidence measures. The local metrics do more than critique; they prioritize. They may reveal that the active site is superbly supported but a regulatory domain is murky, or that a flexible lid over the pocket is the weakest part of the reconstruction. That information can dictate the next experiment. If the lid is functionally important, the lab might collect more data under different conditions, use focused classification to separate states, or add a binding partner that stabilizes the lid. If a particular interface is solid, mutagenesis can be targeted there with confidence. In this way, validation is not only an immune system that removes errors; it is also a nervous system that senses where the model is informative and where it is numb.

Map processing, too, shapes what questions can honestly be asked. Methods such as density modification can improve interpretability by using explicit assumptions about map errors and expected features. When applied carefully, this can be the difference between seeing a ligand clearly and seeing only a suggestion. But the improved clarity is not pure revelation; it is partly inference. The responsible way to use such maps is to treat them as aids for hypothesis generation while anchoring major claims in features that remain stable under conservative processing. In a drug discovery context, for instance, it can be tempting to accept a beautifully shaped ligand density at face value. A more careful approach is to ask whether the ligand placement remains supported across processing choices and whether local confidence measures agree. The questions a structure enables are only as sound as the chain of inference that produced it.

One of the most practically important uses of structure is identifying sites where a protein can be manipulated without destroying it. In protein engineering, the goal is often to change function while preserving fold. An atomic model reveals the protein's load-bearing beams: buried hydrophobic cores, conserved hydrogen-bond networks, and key packing interactions that keep the structure stable. It also reveals the more permissive surfaces where mutations are less likely to collapse the fold. Engineers can design domain swaps, insertions, and surface modifications with a kind of architectural awareness that was previously unavailable. Even when the design fails, it fails in a more informative way, because the structure suggests *why*: a clash here, a lost salt bridge there, a destabilized interface that could be repaired.

Structures also sharpen questions about time, even when the structure itself is a still image. If a channel has a narrow constriction formed by a ring of hydrophobic residues, one can ask whether that constriction is the gate or merely a selectivity filter. If a receptor-binding domain appears in an "up" position in one dataset and "down" in another, one can ask what stabilizes each position and what triggers the shift. The structure suggests experiments that introduce time into the story: capture intermediates by changing temperature, adding ligands, using mutants that slow down transitions, or collecting data quickly after mixing. Cryo-EM's ability to classify particles into different conformations makes these questions especially vivid, because heterogeneity is not merely a nuisance; it is often the phenomenon of interest.

At the human level, there is a particular satisfaction in the way a good structure makes disagreements more productive. Two researchers can argue for years about whether a particular loop is "important." With an atomic model, the argument can become concrete: the loop contacts a ligand, stabilizes an interface, or swings away to expose a pocket. If local confidence metrics say the loop is poorly supported, the argument changes again: perhaps the loop is dynamic, perhaps it adopts multiple states, perhaps it is disordered until a partner binds. In each case, the structure does not eliminate uncertainty; it refines it into better questions. Instead of "is it important?" the question becomes "in which state does it matter, and how can we trap that state?"

The deepest gift of atomic models is not certainty but specificity. A structure does not simply depict a molecule; it proposes a set of levers by which the molecule can be understood and controlled. When those levers are chosen from regions where the map truly supports atomic claims, experiments become sharper, failures become more interpretable, and the distance between a picture and a mechanism shrinks into something a laboratory can traverse with real tools: mutations, ligands, binding assays, and new reconstructions designed to illuminate exactly the part of the machine that still refuses to sit still.

Chapter 6

The Decade the Blurry Blobs Collapsed

For years, cryo-EM promised revelation but often delivered fog: soft-edged densities that hinted at shape without yielding mechanism. Then, in little more than a decade, the fog thinned with startling speed, as new cameras, new mathematics, and new computing habits taught scientists to rescue faint signal from damage, motion, and noise—and to trust what they saw for new reasons.

6.1 When Every Electron Counted Twice

A cryo-EM exposure used to be treated like a single photograph: open the shutter, collect a faint pattern of scattered electrons, and hope the specimen stayed still long enough for the pattern to mean what you thought it meant. Everyone knew it did not. The irony was cruel. The very act of looking—sending electrons through a molecule embedded in glassy ice—nudged the sample. Not much, not in the everyday sense of a hand bumping a table, but enough that the information most precious for atomic-scale interpretation could be smeared into ambiguity.

The shift to movie-mode imaging changed that logic. One exposure became a short film: dozens of frames, each capturing a sliver of time. That sounds like a convenience feature, like swapping an old camera for one with a burst mode. In practice it rewired what counted as data. A frame was no longer a crude fragment of a photograph; it was a timestamped measurement of how the specimen and microscope were behaving during irradiation. The same dose of electrons could now be spent with strategy, not resignation.

To understand why this mattered, it helps to picture what a single particle image really is. Each particle—a ribosome, an ion channel, a viral spike—appears in the microscope as a dim, noisy projection. One particle does not look like a textbook structure. Most of the time it looks like a suggestion. The standard trick, refined over decades, is averaging: if you can align thousands of particles that represent the same object seen in different orientations, the common signal adds up while the random noise cancels away. The catch is alignment. If each particle image is itself blurred—not because the molecule is fuzzy, but because it slid or tilted during the exposure—then you are averaging blur. You may have an ocean of particles and still end up with a map that refuses to sharpen.

Beam-induced motion was the name given to one of the most stubborn sources of that blur. Some of it came from the specimen support flexing; some from charging as electrons deposited energy; some from subtle relaxation of stresses in the frozen layer. Importantly, the motion was not necessarily the same everywhere. Two particles a few hundred nanometres apart might drift differently. A single particle might even move in a slightly curved path rather than a straight line. If you only have one integrated exposure, all of that becomes baked into the image: the particle has traced a tiny smear during the time you were recording it, and the smear steals high-resolution information first.

Movie-mode data collection turned that smear into a track you could measure. Once an exposure is divided into frames, you can ask: where was the particle in frame 1, frame 2, frame 3, and so on? If you can estimate those shifts, you can shift the frames back—computationally—and sum them as if the particle had stayed still. The point is not that the specimen suddenly became well behaved. The point is that its bad behaviour became observable, and therefore correctable.

This is where direct electron detectors earned their reputation. Earlier cameras typically worked by converting electrons into photons in a scintillator and then recording the light; the conversion step spread out the signal and diluted the crispness of each electron event. With direct detection, the detector registers electrons more efficiently, and, crucially, it can read out fast enough that fractionated exposures are practical without turning the measurement into a blur of its own making. Once those detectors were in place, the field began to treat motion correction not as a niche improvement but as a core step in turning raw data into a trustworthy map.

A historical marker often cited in the community is 2013, when Xiao-chen Bai and colleagues demonstrated near-atomic reconstructions using direct detectors and dose-fractionated imaging. The headline result was resolution. The deeper lesson was psychological: what had looked like a hard physical ceiling began to look like a bookkeeping problem. If motion was stealing sharpness, you did not have to accept the theft; you could itemize it.

Imagine filming a dim object through a window while your hands shake. A single long exposure gives you a blurred image. A burst of short exposures gives you many noisy images, but now the blur is not permanent. If you can align the burst, you can recover a composite that is sharper than any individual frame. Cryo-EM movies are like that, except the window is glass ice and the shaking is induced by the beam itself. The signal-to-noise ratio per frame is awful; you cannot simply look at one frame and see the particle clearly. Yet you can still estimate shifts because you have many particles in the field of view, and their collective behaviour provides leverage. Modern motion-correction methods effectively borrow strength from the entire micrograph, tracking coherent drift and also allowing for local differences. A particle's trajectory becomes a best estimate rather than a guess.

Electron counting added a second advantage that is easy to miss if one fixates only on motion. In an integrating camera, a pixel accumulates charge (or light) over a frame, and the result is a continuous value. That is convenient, but it also means the camera is trying to measure a handful of electrons as an analog signal. Counting detectors instead identify individual electron events and register them digitally. For cryo-EM, where the dose must remain low and every electron is precious, this can make the recorded data closer to what you wish you had: a faithful tally of rare events rather than a smeared analog approximation.

The phrase "when every electron counted twice" captures the compounding effect. First, a direct detector made each electron more informative by recording it more efficiently. Second, turning a single exposure into a time series made those electrons more *usable* by letting you undo motion that would have squandered them. It was not that the microscope suddenly produced more electrons. It was that fewer of the electrons you were allowed to use were being wasted on blur.

This new way of thinking encouraged a different style of microscopy. Instead of one exposure per hole in the grid, you now had one exposure *and* a collection of decisions to make about that exposure: how many frames to record; how to align them; whether to treat the whole image as drifting together or allow patches to move differently; how to deal with frames where the specimen has already suffered substantial damage. Those decisions were not cosmetic. They determined which parts of the data were promoted and which were quietly demoted.

Radiation damage does not arrive as a single catastrophic event. It accumulates. Bonds break, charged groups shift, delicate features fade. Even if the specimen remains in place, the image content changes with time because the object itself is being altered. This creates a subtle trap. A later frame might contain roughly the same low-resolution outline as an early frame—the molecule is still broadly there—but the high-frequency information that distinguishes precise shapes is preferentially lost. If you sum all frames equally, you are averaging together better information and worse information as if they were equivalent. That drags the entire result toward the quality of the worst frames.

Dose fractionation makes a different policy possible: exposure weighting. Instead of treating each frame as an equal citizen, you acknowledge that the earliest frames are often the cleanest. They capture the specimen before the beam has had time to do its full mischief. Later frames still carry signal, especially at low and medium resolutions, but they can contribute disproportionately to noise at the very resolutions you care about when you want to interpret chemistry. So you can down-weight later frames, or in some cases discard the most damaged portion of the movie entirely, and still come out ahead.

This is one of those moves that sounds like cheating until you remember what the measurement is for. The goal is not to extract a maximally bright picture; it is to extract the most accurate estimate of the object's structure given a strict dose budget. If the last part of the exposure is mostly paying for damage, insisting on using it "because it is data" is like insisting on listening to a microphone after it has begun to clip and distort, simply because the recording is longer. The time series gives you the power to stop treating damage as an all-or-nothing constraint. Damage becomes something you can model in time and respond to with a choice.

This also changed the emotional texture of a microscope session. In older workflows, a bad micrograph could feel like a sunk cost. You might discover later that drift was worse than you thought, or that some subtle vibration had smeared the data, and there was nothing to do but throw the image away. With movies, "bad" became more nuanced. Perhaps the first few frames were excellent and only later did the specimen lurch. Perhaps the overall drift was large but smooth and therefore correctable. Perhaps the motion was local and patch-based correction would rescue it. The raw data carried clues about its own failure modes.

A concrete example makes the stakes clearer. Consider membrane proteins, long viewed as some of the hardest targets for structural biology. They are flexible, prone to adopting multiple conformations, and often refuse to form the tidy crystals that X-ray crystallography prefers. In 2013, Maofu Liao and colleagues reported a cryo-EM structure of the TRPV1 ion channel at about 3.4 Å. That number was not just a badge. At around that level of detail, the map begins to support specific hypotheses about how a channel might open, how ligands bind, and how mutations alter function. Achieving that kind of map for a mammalian membrane protein depended on a chain of practical victories. Among them was the ability to correct motion and to make smarter use of the limited dose. Movie-mode imaging was not a background convenience; it was part of why the data finally behaved well enough for the biology to become legible.

There is also a quieter, more everyday anecdote that many practitioners recognize. In labs adopting direct detectors for the first time, a common experience was disbelief at the raw movies. The first frame might look unremarkable. Playing the sequence, you could sometimes *see* the specimen shift, not as a dramatic slide but as a steady creep. What had been an abstract source of "blurring" became a visible phenomenon unfolding over fractions of a second. The microscope felt less like a "black box" and more like a system with a pulse.

Once motion correction became routine, new questions emerged. How do you separate true motion from random noise? How do you avoid introducing artifacts by over-correcting? How do you treat cases where the motion is not global but varies across the field? The answers became part of the craft, and eventually part of the software. Tools improved, culminating in widely adopted approaches such as MotionCor2 (Zheng and colleagues, 2017), which treated motion not as a single shift but as something that can vary across the image. The message for non-specialists is not the tool name, but what it represents: the maturation of a capability from "possible in principle" to "reliable enough that many labs can use it without heroics."

The combination of motion correction and exposure weighting also changed what "resolution" meant in practice. Before, resolution improvements often relied on gathering more particles or hoping for a better specimen. Now, some of the improvement came from getting more truth out of the same particles. A dataset could be reprocessed and suddenly yield sharper features, not because the microscope had been different, but because the time dimension had been used more intelligently. That is an unusual kind of progress: it is not a new lens, but a new way of spending the same photons—or, here, the same electrons.

There is an almost philosophical twist hiding in this technical story. In many sciences, measurement is about freezing the world: capturing an instant, isolating a variable, pinning down a value. Cryo-EM, under extreme low-dose constraints, learned to do something closer to accounting. The specimen changes as you observe it; the act of observing causes motion; information arrives in a sequence that decays in quality. Movie-mode imaging does not deny those facts. It embraces them. It says: if the measurement is inherently time-dependent, then time should not be thrown away.

That recognition also influenced how people talked about the data. Micrographs stopped being purely images and started being records. A movie contains not just a view of particles but a history of what happened during irradiation: how quickly drift settled, when damage began to dominate, whether charging was severe. In the hands of an experienced microscopist, that history can guide the next steps at the microscope: changes in support films, beam settings, grid quality, or collection strategy. In other words, the time series is not only a way to improve a reconstruction later; it is a diagnostic instrument in its own right.

The end result can look deceptively simple: a sharper particle average, a cleaner map, a structure that supports confident interpretation. Yet the path to that clarity runs through an admission of fragility. Biological specimens under an electron beam are not stable targets; they are perishable. The breakthrough was not to pretend otherwise, but to build a workflow that extracts the earliest, least compromised information and then uses computation to correct what can be corrected. The harsh limit—you can only afford so many electrons before the specimen is altered—did not disappear. It became a constraint you could plan around, frame by frame, electron by electron, making the scarce signal do more work than it had ever done before.

6.2 The Maps That Refused Easy Answers

The new detectors and movies did not merely sharpen cryo-EM; they raised the stakes. Once the data began to carry more real structural signal, it became harder to excuse a map that was wrong. A blurry result can be ambiguous in a forgiving way: a bump could be a helix, or a loop, or noise, and nobody can quite prove you mistaken. A sharper map has sharper liabilities. If you claim to see a particular arrangement of protein strands or a ligand tucked into a pocket, other scientists will build explanations on top of it. The field needed something beyond clearer images: it needed a way to keep itself honest when the temptation to over-interpret was stronger than ever.

A single-particle reconstruction is, at heart, an inference problem carried out under duress. You never observe the molecule directly. You observe thousands or millions of noisy projections, each one a snapshot of one molecule in one unknown orientation, embedded in glassy ice. The computer's job is to align these projections and combine them into a three-dimensional map. The danger is that alignment is not simply reading off a value; it is making a choice. If the image is noisy, many orientations can seem plausible. If the molecule itself is flexible or partially occupied by binding partners, more than one answer may truly be correct. Older approaches often behaved as though there must be one best answer and the algorithm's duty was to find it. That worked tolerably when maps remained vague. As soon as cryo-EM began promising something closer to an atomic portrait, that attitude became risky.

The new generation of software grew up with a different temperament: a cultivated reluctance to commit too early. Instead of asking, for each particle, "what is the orientation?", the software began to ask, "which orientations are *plausible*, and how confident are we?" Instead of insisting that every particle belongs to the same perfect class, it allowed that some particles may be junk, some may be damaged, some may capture the molecule mid-wiggle, and some may represent a different biochemical state entirely. The revolution was as much psychological as mathematical. Cryo-EM became better by learning to distrust its own first impressions.

One of the most visible manifestations of this skepticism was classification. The word sounds bureaucratic, but the idea is intuitive. If you photograph a crowd from far away and then try to average all the faces into one, you get a blur. If you first sort the crowd into smaller groups—roughly similar poses, roughly similar faces—each average becomes clearer. In cryo-EM, 2D classification groups particle images into families of similar-looking projections. Some families correspond to genuine views of the target molecule. Others correspond to contamination, ice crystals, broken particles, or complexes that fell apart. Sorting is not glamorous, but it is how a reconstruction avoids being dragged down by images that never had the desired signal to begin with.

Early cryo-EM practitioners did classification, but it was often fragile, manual, and prone to wishful thinking. It is surprisingly easy to "recognize" what you hope to see in noise. Anyone who has stared at clouds and found animals has the relevant brain wiring. The field's cautionary parable has a name: "Einstein from noise." Long before cryo-EM's resolution revolution, image-processing demonstrations showed that if you align and average pure noise while nudging the alignment toward a desired template, the template can emerge as if by magic. The result is compelling to the eye and utterly false. The moral is not that scientists are foolish, but that pattern-seeking is automatic, and computers will obediently amplify it if asked.

Model bias is the more formal name for the same hazard. If the algorithm begins from an initial guess—a low-resolution map, a related structure, a hand-drawn model—the refinement procedure can be pulled toward that guess. Sometimes that pull is appropriate; an initial model can help a difficult problem find its footing.

Sometimes it becomes a hidden steering wheel. The new software culture did not ban initial models, but it treated them with suspicion, insisting on checks that ask whether the data genuinely support the answer or are merely being coerced into it.

That is where the probabilistic turn mattered. Bayesian refinement, popularized in cryo-EM by Sjors Scheres with RELION (2012), rephrased the central task in the language of probabilities: given a 3D map, how likely are the observed images? Given noisy images, how much support is there for different orientations, different alignments, different class assignments? In practice, this meant the algorithm could distribute its bets. A particle image might contribute a little evidence to one orientation and a little to another, rather than being forced into a single bin early on. As refinement proceeds and the map improves, those probabilities sharpen; uncertainty shrinks where the data are informative. Where the data remain ambiguous, the software does not pretend otherwise. It carries uncertainty forward instead of sweeping it under the rug.

This approach also introduced a kind of discipline that non-specialists might recognize from everyday decision-making. If you are trying to identify a person in a grainy security video, you might begin with several candidates. As you gather more evidence, you narrow the list. What you should not do is decide immediately and then interpret every pixel as confirmation. Bayesian refinement encodes that commonsense rule in mathematics. It builds an incentive to avoid overconfident alignment choices when the evidence is weak, because those choices can later harden into artifacts.

The benefits were not merely philosophical. They were practical and measurable. When particle alignments become more reliable, you need fewer particles to reach a given level of detail, because you are no longer averaging together images that are subtly mis-registered. Bai and colleagues' 2013 demonstrations of near-atomic reconstructions did not succeed only because the detector delivered cleaner images. They also relied on better alignment and refinement practices that could extract the additional signal without corrupting it. In other words, the hardware delivered opportunity; the algorithms determined how much of that opportunity was actually collected.

The same skeptical spirit drove the field's insistence on anti-overfitting safeguards. As already discussed, gold-standard independent half-map refinement, articulated clearly by Scheres and Shaoxia Chen (2012), became a widely accepted remedy for a particular self-deception: if you use the same data both to refine a map and to judge its quality, you can fool yourself into thinking the map is sharper than it truly is. The deeper significance is cultural. Cryo-EM's credibility depends not on a single resolution number but on workflows designed to catch human and computational enthusiasm before it outruns the evidence.

That brings up another complication that the "one true map" mindset tended to hide: biological molecules are often not single objects at all, but ensembles. The classic single-particle assumption—that every particle is an identical copy—is an approximation. Sometimes it is a good one, as with symmetric viruses or rigid enzymes. Often it is only partly true. A membrane receptor might have a stable core and a flexible tail that swings through many positions. A complex might lose a subunit in a fraction of particles. A channel might be half in one conformation and half in another, because the grid captured it mid-transition. Treating such mixtures as if they were uniform does not merely blur the result; it can generate a chimera, a map that averages incompatible states into a physically misleading compromise.

Classification extended from 2D to 3D to address this. In 3D classification, the algorithm sorts particles into groups that support different three-dimensional maps. One class might represent a closed conformation; another, an open one. One might include a bound ligand; another might not. The aim is not to multiply answers for its own sake, but to prevent incompatible data from being mashed into a single map that fits nothing well. This is where the phrase "the maps that refused easy answers" earns its meaning. The best software began to insist that the data were saying, gently but firmly, "there is more than one story here."

A concrete illustration comes from targets like γ-secretase, a membrane-embedded complex associated with Alzheimer's disease biology. When Bai, Scheres, and colleagues reported an atomic structure of human γ-secretase (2015), part of the achievement was not only reaching around 3.4 Å but negotiating heterogeneity without surrendering interpretability. Large membrane complexes are not static sculptures. They breathe, flex, and sometimes partially disassemble. If your processing pipeline assumes perfect uniformity, it will punish you for that biological reality. Heterogeneity-aware classification and refinement made it possible to keep the stable parts sharp while acknowledging where the molecule was less cooperative.

A related insight, now almost routine, is that resolution is not uniform across a map. In the older era, a single global number often stood in for quality, as if a map were equally reliable everywhere. In practice, near-atomic cryo-EM maps can be uneven: flexible domains, partial occupancy, compositional heterogeneity, and preferred orientation lead to locally variable detail rather than uniformly crisp density. Local resolution mapping, formalized by tools such as ResMap, made this unevenness visible and discussable. The point is not to embarrass a structure by revealing its weak spots. The point is to avoid the easy lie of uniform certainty.

Consider what this means for a biologist reading a structure. If the core of a complex is well resolved but a peripheral domain is blurred, you should not model both with the same confidence. A careful structural biologist will build an atomic model where the map truly supports it and will resist inventing precise coordinates for regions that are effectively a fog. Local resolution estimates provide a language for that restraint. They allow authors to say, with specificity, "this region supports a detailed model; that region should be interpreted as mobile or uncertain." This turns what might once have been dismissed as imperfection into biological information. Flexibility becomes not merely a nuisance but a feature that can be mapped.

Preferred orientation is another source of discomfort that smarter algorithms helped manage. If particles land in the ice in mostly one or a few orientations, the dataset contains an imbalanced set of views. This makes some directions in 3D space well supported and others poorly supported, and the reconstruction may develop anisotropic detail: sharp in one direction, smeared in another. A naïve refinement can mistake these imbalances for genuine features, or it can overfit the poorly supported directions by amplifying noise. Classification, careful refinement, and diagnostic tools can reveal the problem early enough that a scientist might adjust sample preparation or data-collection strategy. Again the theme is skepticism: not treating the first reconstruction as an oracle, but as a hypothesis that must be tested against the known ways cryo-EM can mislead.

All this skepticism might sound like it slows science down. In practice it did the opposite, because it reduced the number of dead ends. A map that is slightly too good to be true is worse than a map that is openly uncertain. The former invites years of misguided work; the latter invites better experiments. The modern cryo-EM mindset prizes reproducibility and robustness: rerunning refinements, comparing outcomes across processing choices, and documenting the path from raw images to final map. Validation discussions increasingly treat trustworthiness as an outcome of workflow design—splits, controls, metrics—rather than just map appearance. A beautiful map is not enough if the road to it was paved with hidden assumptions.

There is also an emotional shift here, one familiar to anyone who has watched a field mature. Early on, progress is fueled by optimism: the conviction that if you push hard enough, a clear answer will emerge. Later, confidence must be earned differently. The question becomes not “can I get a structure?” but “can I justify each interpretive step to someone skeptical and smart?” Cryo-EM's decade of transformation included that change in temperament. It was not only a contest against physics and instrumentation, but a contest against our own eagerness to see patterns and declare victory.

What makes this story compelling is that the reward for skepticism was not mere caution but new kinds of insight. Once algorithms stopped forcing particles into a single best answer, they could reveal mixtures of states that had been invisible as anything other than blur. Once refinement routines stopped pretending every alignment was certain, they could use uncertainty itself as information, gently steering the reconstruction toward what the data could actually support. And once validation became part of the everyday workflow rather than an afterthought, cryo-EM maps became objects that other scientists could trust enough to act on: to design mutations, to interpret disease variants, to propose mechanisms, to build drugs.

A good cryo-EM map, in this modern sense, is not a picture that shouts. It is a piece of evidence that has learned how to speak with the appropriate level of confidence: firm where the data are strong, quiet where the data are thin, and never pretending that biology must be simpler than it is.

6.3 Speed Became a Scientific Method

For a long time, cryo-EM had an awkward rhythm. A lab might spend a precious night on the microscope collecting images—the expensive part, the part scheduled weeks in advance—and then spend weeks waiting to learn whether those images contained a usable structure. The delay was not just inconvenient. It encouraged a certain kind of scientific gambling: collect as much as possible, hope for the best, and only later discover what you should have done differently. When the answer arrives a month after the experiment, you cannot go back in time and adjust the ice thickness, the buffer, the particle concentration, the microscope alignment, or the data-collection strategy while you still have the grid in the holder and the beam tuned.

The change in the 2010s was not simply that computers got faster. Speed began to shape behaviour. Once processing could keep pace with data collection, cryo-EM stopped being a single long bet and became a series of short, testable decisions. The experiment no longer ended when the microscope session ended. It continued, in near real time, as computation fed back into what the microscopist chose to do next.

To see why speed mattered so deeply, it helps to remember how many moving parts a modern single-particle workflow contains. A dataset is not a single object you either have or do not have. It is a chain of transformations: raw movies become aligned micrographs; micrographs produce particle images; particles are sorted into classes; a 3D map is built and refined; validation checks are run; alternative processing choices are compared. Each step has parameters and judgement calls. A little change early can ripple forward and change the outcome dramatically. When the whole pipeline takes weeks, those parameters become sticky. People tend to commit early, because revisiting decisions is too expensive. When the pipeline takes hours, revisiting becomes normal, and the quality of science rises with the number of times you can ask, calmly, *is this robust?*

The most obvious engine behind the new tempo was the GPU, the graphics processing unit. GPUs were originally designed for rendering images and video games, where you must perform the same operation on many pixels at once. Cryo-EM happens to have a similar shape: you perform related computations on vast numbers of particles, frames, and Fourier-space pixels. A GPU is not magically smarter than a CPU, the ordinary computer processor; it is simply built for parallelism. Instead of doing one calculation at a time very quickly, it does many calculations at once. Cryo-EM algorithms, once adapted to that style of computing, could run orders of magnitude faster in the steps that dominate time: alignment searches, refinement iterations, and classification.

Ali Punjani and John Rubinstein's cryoSPARC (2017) became a symbol of this shift because it made speed visible to everyday users. A task that once demanded a queue on a cluster and a weekend of waiting could sometimes be explored interactively, in something closer to an afternoon's work. The details of the algorithm matter to specialists, but the broader impact was experiential. Suddenly, a student could try a different particle-picking threshold, rerun a 2D classification, and see the consequences without turning it into a week-long project. A postdoc could test whether an observed feature survived a change in masking or refinement settings and know the answer before the next lab meeting. A principal investigator could ask for an additional control refinement without feeling like they were requesting a moon landing.

Speed changed what counted as a reasonable question. When processing is slow, you ask "can we get *a* reconstruction?" When processing is fast, you ask "how sensitive is this reconstruction to our choices?" That is a different kind of ambition. It is less about arriving at a picture and more about arriving at a claim you can defend.

The software culture that emerged around RELION contributed to the same behavioural change from another angle. RELION's automated refinement workflows, including "3D auto-refine" (Scheres, 2012), reduced the amount of manual tuning and idiosyncratic decision-making that had quietly shaped results. In older practice, an experienced user might coax a dataset into a good-looking map through a sequence of small interventions: adjust a parameter here, modify a mask there, restart an iteration, smooth an angular distribution. Some of those interventions were legitimate expertise. Some were dangerously close to "expert hand-tuning" as a hidden variable. If a result depends on a particular person's tacit knowledge, it is harder for the community to reproduce and harder for the field to trust.

Automation, in this sense, was not about removing humans from science. It was about making the human contribution explicit and repeatable. A workflow that can be rerun by someone else, with recorded settings and systematic validation, is a workflow that can support a shared evidence standard. Speed and automation reinforced each other: once a run finishes quickly, it becomes feasible to rerun it from scratch instead of making ad hoc midstream edits that are hard to document.

The less celebrated but equally important part of the tempo shift was the rise of pipelines: software ecosystems that treat data processing as a managed process rather than a craft performed by memory and handwritten notes. Two names capture this ethos from an earlier phase. Leginon, designed for automated data acquisition, helped make it possible to collect large datasets systematically rather than by hand, image by image. Appion, focused on processing provenance, treated the analysis not as a set of isolated files but as a recorded history of decisions. In Appion's spirit, the result is not merely "a map"; it is a map with a traceable lineage: which micrographs were used, which particles were discarded, which settings produced a particular classification. These "computing habits" might sound administrative, but they are what allow large teams to collaborate and allow a lab to understand its own past choices months later.

At first glance, data acquisition automation and processing provenance seem separate from GPU acceleration. One is about the microscope; the other about the computer. In practice they converged into a new way of working: collect more data, process it on the fly, and let the results guide what happens next. The microscope became less like a camera and more like an instrument in a loop, where measurement and interpretation happen in the same time window.

A concrete modern scenario makes this vivid. A team begins a microscope session with a new protein complex. The sample is precious; perhaps it took months to purify and behaves differently each time. They load a grid and start collecting. Within the first hour they might have a few hundred micrographs. In the older era, those images would be copied to storage and then left untouched until the session ended. Now, as the microscope continues collecting, the first batch is already being processed. Movies are aligned; micrographs are inspected; particles are picked and run through quick 2D classification. A first attempt at an *ab initio* 3D model—a reconstruction built without a detailed prior model—can be launched. By the time the night is half over, the team may already know whether the particles are intact, whether preferred orientation is severe, whether aggregation dominates, whether multiple states are present, or whether the dataset is simply too messy to justify another eight hours of collection.

The psychological effect is hard to overstate. A microscope night used to end with uncertainty: exhausted people walking away with terabytes and hope. Now it can end with a decision: continue because the data are promising, or stop because the sample or grid conditions need to be changed. That is not merely efficiency. It is a way of respecting scarce resources: microscope time, reagent costs, and the human time of a team.

This feedback loop depends on another kind of "speed" that is not purely computational. Standardized tooling reduces the number of fragile, bespoke steps that once slowed labs down and made outcomes brittle. Motion correction is a good example. Once motion correction became reliable and widely deployable—with tools such as MotionCor2 (Zheng and colleagues, 2017) turning what had been a specialist procedure into a more standardized one—the pipeline became smoother and less dependent on a particular expert. When a step becomes routine, it stops being a bottleneck. It also stops being a source of mysterious differences between labs.

Speed also made validation more than a slogan. In principle, rigorous scientists always want to test robustness: run alternative refinements, change a mask, vary a classification parameter, compare outcomes. In practice, those tests were often skipped because they were too time-consuming. Once GPUs and streamlined software made reruns cheap, validation became a normal part of the day's work rather than an aspirational appendix. If a structure changes substantially when you adjust a reasonable parameter, that is a warning sign worth noticing. If it remains stable across multiple choices, confidence becomes something you can earn rather than simply feel.

There is a revealing historical anecdote, repeated in different forms across labs, about the arrival of this new tempo. People describe the moment they realized that computation had stopped being the slowest part of the process. In the earlier period, it was common to launch a refinement and then plan the rest of the week around it: check progress tomorrow; hope it converges by Friday. The algorithm's pace set the lab's pace. With GPU acceleration, that relationship inverted. Refinements became something you might start after lunch and inspect before dinner. That changed how people interacted with their own data. They became more inquisitive and less superstitious, because they could test ideas quickly rather than protecting a single processing run as if it were fragile.

This shift also altered training. In a slower era, the craft of cryo-EM analysis could feel like apprenticeship: you learned by watching an expert make choices you could not easily justify from first principles. In a faster era, newcomers can learn by experimentation. They can run the same dataset through two workflows and compare outcomes, not as a theoretical exercise but as something they can do on a workstation. That kind of learning is often more durable. It builds intuition grounded in observation rather than lore.

Not all consequences of speed were automatically positive. Faster iteration can encourage sloppy iteration: the temptation to keep tweaking until you get a pleasing map, like adjusting photo filters until an image looks right. The field's increasing emphasis on documented workflows and validation was partly a response to that temptation. When speed rises, the ethical burden of restraint rises too. Fortunately, the same computational power that enables endless tweaking also enables systematic controls. A lab can decide to rerun refinements with different random seeds, to test multiple plausible masks, to split data in different ways, and to store a record of those choices. Speed can support discipline, but only if the culture insists on it.

The real transformation was that cryo-EM became less of a one-shot performance and more of an iterative practice. In the old rhythm, the microscope session was the performance and the computation was the critic writing a review weeks later. In the new rhythm, computation sits in the room. It speaks up early. It tells you when the sample is degrading, when the data are dominated by junk, when the particles have adopted a single view, when the reconstruction is stable, when it is not. It turns the experiment into a conversation between the physical specimen and the analysis, happening on the timescale of hours rather than seasons.

By the end of the decade, it was no longer unusual for a lab to treat a cryo-EM project as something that can be steered day by day. The faster computer did not simply make old workflows quicker. It made it sensible to ask better questions, to demand more checks, to treat uncertainty as something you can measure, and to let evidence, not optimism, decide when to press on and when to change course.

6.4 The Day Cryo-EM Stopped Being a Gamble

A scientific technique becomes ordinary in a particular way: it starts to disappear into other people's plans. Grant proposals stop treating it as a daring add-on and begin to list it as a routine route to an answer. Students expect to be trained on it. Reviewers begin to ask why it *wasn't* used. That is what "mainstream" means in practice. It is not a press release or a prize. It is the moment when a method becomes reliable enough that strangers can build their work on your result.

Cryo-EM reached that point in the 2010s because several kinds of reliability arrived together. Better detectors and better algorithms made sharper maps feasible, but feasibility alone does not make a technique widely trusted. What changed the field was the emergence of shared habits: public software that many groups could run, standard validation practices that made overconfidence harder, benchmark specimens that allowed apples-to-apples comparison, and a small set of unmistakable success stories that made it emotionally safe for biologists outside the specialist community to invest.

Trust starts with something simple: the same dataset should not produce wildly different structures in different hands. In the era when cryo-EM often yielded "blobs," that demand was difficult to meet, partly because the data were weak and partly because processing was artisanal. Two labs could make different choices about particle selection, alignment, or filtering and end up with maps that had the same vague silhouette but different implied mechanisms. The argument would then move from evidence to authority: whose judgement was better, whose experience deeper. Such debates are not unique to cryo-EM, but they are a warning sign of a method that has not yet stabilized.

Public, actively maintained software was one of the quiet stabilizers. RELION, introduced by Sjors Scheres (2012), was important not only for its statistical framing but because it offered a coherent, widely adoptable workflow. It nudged the community toward running comparable procedures instead of each group improvising. The value of that is hard to appreciate until it is absent. When a field shares tools, it shares failure modes too. Everyone learns, collectively, which steps are fragile and which are robust. Everyone can test an improvement against a common baseline. "I can reproduce your result" stops being a compliment and becomes a default expectation.

The parallel development of cryoSPARC (Punjani and Rubinstein, 2017) reinforced this ecosystem effect. Cryo-EM began to have not one "house style" but a small number of broadly used toolchains. That diversity mattered. If two independent software suites can process the same particles and converge on the same structural conclusion, confidence increases. Disagreement is also productive: it forces people to ask whether a difference comes from a genuine biological ambiguity or from a software assumption. Either way, the conversation becomes more concrete.

A second stabilizer was validation culture, particularly the adoption of gold-standard anti-overfitting practices. The change was not merely technical; it was moral. When the community began treating independent half-map refinement as a norm rather than an optional extra, it drew a line between "a map that looks persuasive" and "a map that has survived a specific test designed to catch self-deception." Scheres and Shaoxia Chen's 2012 work on preventing overfitting crystallized this as a procedural expectation. The field's claims became more believable not only because maps improved, but because the controls improved along with them. This is how trust grows in a method: not by insisting that artifacts are rare, but by designing workflows that make artifacts easier to detect.

Benchmark specimens helped in a different way. In any measurement technology, shared targets act like calibration weights. If dozens of labs can solve the same sample and report consistent outcomes, then differences on new samples are more likely to reflect biology rather than idiosyncrasy. Structural biology has long relied on this kind of informal calibration. In cryo-EM, the need was especially acute because the pipeline contains so many choices. Benchmarking turned arguments into experiments: if a new algorithm truly helped, it should help on a known test case; if a new detector improved real information capture, the improvement should appear across many groups, not just in a single heroic lab.

Still, communal habits alone do not create a tipping point. People are persuaded by stories, and science is no exception. Cryo-EM needed structures that made outsiders stop thinking of it as a specialist's last resort and begin thinking of it as an attractive first option. Three milestones, arriving in quick succession, played that role with unusual force.

The first was TRPV1. In 2013, Maofu Liao and colleagues reported a structure of the TRPV1 ion channel at about 3.4 Å. Ion channels are membrane proteins, embedded in lipid bilayers, often flexible, and historically difficult. For decades, "membrane protein structure" had been almost synonymous with heroic crystallography campaigns: long timelines, high failure rates, and frequent compromises. TRPV1 signaled a break with that narrative. Here was a mammalian membrane channel solved by cryo-EM without crystallization, in a range of detail where chemical interpretation becomes plausible. The accomplishment did not merely impress cryo-EM practitioners; it changed the mental math for biologists who had spent years watching membrane targets defeat other techniques.

The second milestone was biomedical seriousness, embodied by γ-secretase. In 2015, Bai, Scheres, and colleagues reported an atomic structure of human γ-secretase at roughly 3.4 Å. The complex is multi-subunit and clinically relevant, tied to Alzheimer's disease biology. It also carries the kind of complexity that often reveals the limitations of a method: heterogeneity, flexible regions, partial occupancy. Producing an interpretable structure of such a target made cryo-EM feel less like a boutique technology and more like a practical tool for problems where the stakes are high and the molecules refuse to cooperate.

The third milestone pushed the notion of precision. In 2015, Alberto Bartesaghi, Alan Merk, and colleagues reported a cryo-EM structure of β-galactosidase in complex with a cell-permeant inhibitor at about 2.2 Å. At that level, the conversation shifts. The question is no longer "can you trace the backbone?" but "can you model chemistry with confidence?" A structure at $\sim$ 2 Å invites claims about inhibitor interactions and subtle rearrangements. It also forces the community to confront validation even more sharply, because the temptation to over-interpret grows as maps become more detailed. In that sense, β-galactosidase did two things at once: it persuaded skeptics that cryo-EM could reach a crystallography-adjacent regime for suitable samples, and it raised the standards for how rigorously such claims must be checked.

These successes were not isolated fireworks. They arrived amid a broader change in expectations that Werner Kühlbrandt captured memorably in 2014 by calling it the "resolution revolution." The phrase stuck because it named a social reality, not just a technical one. A field changes when a result goes from being a headline to being something students are expected to produce in their training. Cryo-EM began to generate structures often enough, and with enough interpretability, that the community's baseline shifted. The revolution was not that every molecule suddenly became solvable. It was that many classes of molecule moved from "possible if you are lucky" to "reasonable to attempt if you do competent work."

A method becomes a gamble when success depends heavily on the personality of the lab: a particular virtuoso microscopist, a particular software whisperer, a particular set of undocumented tricks. Cryo-EM stopped being a gamble as the hidden variables were squeezed out. Automated refinement reduced dependence on hand-tuning. Standardized tooling reduced dependence on custom scripts. Pipelines and provenance tracking reduced dependence on memory. Training materials, workshops, and shared datasets made it easier for newcomers to learn what "good practice" looks like. When many groups can converge on similar answers, the method begins to belong to the community rather than to a few specialists.

One can see this shift in the way collaborations changed. In the earlier era, a cryo-EM collaboration might be framed apologetically: "we tried crystallography and NMR; they failed; perhaps cryo-EM can at least tell us the overall shape." As confidence grew, the same kind of collaboration could start differently: "we have a functional question and a difficult target; cryo-EM may be the fastest path to a mechanism." That shift affected not only what was studied but how quickly ideas moved. A structural result could now appear early enough in a project to guide experiments, rather than arriving as a retrospective explanation.

The mainstreaming of cryo-EM also altered the relationship between structure and uncertainty. Earlier, ambiguity was often an embarrassment: the map was too soft in places, too uneven, too hard to interpret cleanly. Later, the community developed a more mature stance. Uneven maps became discussable. Heterogeneity became something to analyze rather than to hide. Validation became something to present as part of the evidence rather than as an administrative hurdle. This did not make the structures less human; it made them more honest. In biology, flexibility and mixtures of states are often the rule rather than the exception. A method that can represent that reality, while also signalling where its own certainty ends, is a method that other scientists can use responsibly.

The day cryo-EM stopped being a gamble was not a date on a calendar. It was the accumulation of ordinary successes achieved by ordinary labs, backed by workflows that were becoming standardized enough to teach, and guarded by checks that made exaggerated claims easier to spot. When a method reaches that point, its most important products are no longer individual structures, impressive as they may be. Its most important product is confidence shared across a community: the sense that if you prepare a good sample, collect thoughtful data, and process it with disciplined controls, the map you obtain will be solid enough for other people to build their own science on top of it.

Chapter 7

Three Characters and a Frozen Epiphany

Cryo-EM did not become transformative because one machine suddenly got better; it became transformative because a few stubborn thinkers solved different pieces of the same impossible puzzle. Behind the celebrated images lies a drama of ruined samples, faint signals, mathematical rescue, and a final shift in belief: that biology's smallest machines could at last be seen almost atom by atom.

7.1 When Water Refused to Behave

Electron microscopes had always offered a kind of cruel promise to biologists. The physics was generous: electrons, with their extremely short wavelengths, could in principle resolve details far smaller than light microscopes could ever manage. The biology was fragile. A cell, a virus, a protein complex in solution was mostly water, and water was the one thing the electron microscope was least willing to tolerate. In the high vacuum of a transmission electron microscope, liquid water would boil away. And if you tried to outsmart the vacuum by freezing the sample, ordinary ice turned out to be a saboteur.

It is hard to convey to a non-specialist how demoralizing that was.

Imagine having a camera capable of photographing a hummingbird's feathers one by one, but every time you raised it to your eye, the air itself crystallized into needles that shredded the bird. For years, biological electron microscopy lived with compromises. There were ways to remove water chemically and replace it with plastic, or to stain the sample with heavy atoms so that the outlines became visible. Those methods created powerful images and founded entire areas of cell biology. But they also made an implicit bargain: you would see something, not necessarily *the thing as it was*.

What, exactly, was wrong with ordinary freezing? Ice is not just "water but solid." When water freezes slowly enough for its molecules to find their preferred arrangement, it forms crystals: an orderly lattice with sharp boundaries between growing grains. That order is the problem. As crystals form, they push solutes aside. They expand with angles and facets. Their growth is mechanical as well as chemical. In a thin film containing proteins, membranes, or whole viruses, the advancing front of ice crystals can shove objects out of place, squeeze them, or tear delicate assemblies apart. Even if the specimen survives the squeeze, crystalline ice becomes an intrusive foreground: it scatters electrons strongly and in patterned ways, adding its own regular "signal" that can overwhelm the subtle, irregular information a biologist is hoping to measure.

The result was a paradox. To make the sample compatible with the vacuum, you froze it. But freezing, done naively, changed the very structure you cared about. The electron microscope could be a perfectly good instrument pointed at a specimen that was no longer trustworthy.

Jacques Dubochet's contribution was to insist that this wasn't a philosophical limitation, but a practical one: perhaps water did not always have to crystallize. Perhaps, under the right conditions, it could be trapped in a solid state without making crystals at all. That idea had a long prehistory in physics and physical chemistry. Many liquids can become glassy solids if you cool them fast enough that molecules have no time to organize into a crystal lattice. But water is notoriously eager to crystallize; it is the substance that makes the winter world look so confidently geometric. To turn water into a glass required treating it not as a background medium but as an active, temperamental material with its own kinetics, its own demands about thickness and speed.

Dubochet came to the problem with a stubborn clarity about what a "usable specimen" should mean. A usable specimen was not one that merely survived the microscope. It was one that preserved the native arrangement of biological molecules closely enough that the resulting image could be interpreted as biology, not as the aftermath of preparation.

Two constraints made the problem more brutal and more tractable. The brutal constraint was speed. To avoid crystal formation, the sample had to be cooled so rapidly that water molecules were immobilized before they could line up. The tractable constraint was thickness. Cooling rate is not an abstract number; it depends on how quickly heat can leave the material. A thick droplet cools from the outside in, leaving time for crystal growth in the interior. A very thin film can be quenched almost instantly. Biological electron microscopy already dealt in thinness because electrons cannot penetrate deeply. In other words, the specimen had to be thin for imaging anyway. Dubochet's insight was that thinness was also the key to turning water into an invisible support rather than a destructive crystal factory.

The practical theatre for this is the electron microscopy grid: a small metal mesh that holds an ultrathin support film across its holes. Onto that support, a tiny droplet of sample is applied. The droplet is not what you ultimately want; it is far too thick. It must be blotted—touched with filter paper so that most of the liquid is wicked away—leaving a film of water spanning the holes, often only tens of nanometres thick. Within that film, the molecules are suspended in their native buffer. In older approaches, one might then let the water evaporate or chemically fix the sample. Dubochet's approach was to freeze that film so quickly that it became glass.

This is where the method becomes both dramatic and strangely domestic. The decisive move is a plunge: the grid, held in tweezers, is dropped rapidly into a cryogen. Not simply into liquid nitrogen, though nitrogen is cold enough to make the air sting and metal burn. Liquid nitrogen boils violently when it touches something warm, creating a cushion of gas that actually insulates the sample and slows cooling at the surface. A better choice is a cryogenic liquid such as liquid ethane cooled near its freezing point by liquid nitrogen. Ethane wets the grid more effectively and boils less turbulently, pulling heat out fast. The difference is not a minor refinement; it is the difference between water having time to become a crystal and water being caught mid-motion, fixed into a glassy arrangement.

The early years of this work had the texture of tinkering that becomes, in hindsight, an intellectual pivot. In his later reminiscence (2016), Dubochet describes the push to make vitreous water not a heroic trick but a routine fact. Routine is an underrated word in scientific revolutions. A method only changes a field when it can be repeated by other hands, in other rooms, on ordinary days. A single beautiful demonstration can be admired and then quietly bypassed. A routine method becomes infrastructure: it changes what questions people dare to ask.

One can see why the skepticism was persistent. Water is a small molecule, a champion at forming crystals. Biological samples are complex: salts, glycerol, proteins, lipids, nucleic acids. The film must be thin but not too thin; it must not dry; it must not pick up contamination; it must be transferred into the microscope without warming enough to crystallize. Each step seems like an invitation for failure. Yet it was precisely by treating each step as negotiable—by learning what the water was likely to do and then forcing it to do something else—that Dubochet and colleagues made the technique workable. The 1980s papers culminating in Dubochet and Adrian's 1988 account did more than report a trick; they established a new specimen state as a credible standard: biological molecules embedded in vitreous ice, close to their aqueous, native arrangement.

A concrete example makes the transformation easier to appreciate. Consider a virus particle, a small protein shell surrounding genetic material. In solution it is a robust object, but it is also defined by delicate, repeated protein interactions, the sort that can be subtly distorted by dehydration or salt concentration changes. A slowly frozen droplet can deform such particles, and crystalline ice can impose strong patterns in the image that distract from the viral features. In a vitrified film, the particle is held in place without being pressed against a surface or surrounded by growing crystals. The ice itself becomes almost conceptually absent: it is there to keep everything where it belongs and to provide a background of weak, featureless scattering. What remains to be seen is the virus.

Or consider a large molecular machine such as the ribosome. It is not just a blob; it is a complicated assembly with moving parts, and its function depends on its shape and its flexibility. Chemical fixation and heavy staining can lock it into unnatural conformations or emphasize some regions at the expense of others. Vitrification does not preserve motion—freezing halts movement—but it can preserve *a moment* without demanding that the molecule be flattened, dried, or chemically reinforced. You trade time for truth: a snapshot instead of a sculpture.

The most subtle part of Dubochet's shift was psychological. Before vitrification, the specimen was often treated as a reluctant victim: something to be coerced into a vacuum-compatible form by any means that yielded contrast. After vitrification, the specimen became something more like a preserved state: the goal was not simply to see *something* but to see *the right thing*. That change in ambition altered what counted as progress. Better microscopes mattered, but better specimens mattered first.

There is also a quiet elegance in the physical logic. A vitrified film solves multiple problems at once. It prevents evaporation in the vacuum. It suppresses the violent rearrangements of crystallization. It immobilizes molecules so that the microscope can image them. It reduces the need for heavy-metal stains that can obscure fine features. It also creates a reproducible geometry: a thin layer spanning a hole in a support film, a flat landscape where many identical particles can be found in random orientations. That last point would become decisive for later computational methods, but even before those methods matured, the geometric simplicity was invaluable. It gave the microscopist something stable to work with.

Stability, however, is never absolute. Vitrified specimens remain vulnerable in their own way. Warm them slightly and they can devitrify—transition from glass into crystals—destroying the very state you have worked to trap. Expose them too long to the electron beam and the molecules will change chemically and physically. Handle them carelessly and they can pick up frost, a fine contamination that looks innocuous to the eye but becomes ruinous under the microscope. The point is not that Dubochet eliminated the difficulty; it is that he moved the difficulty into a domain where disciplined technique could manage it. Instead of the sample being fundamentally incompatible with the instrument, the sample became conditionally compatible, provided the experimenter respected the rules of glassy water.

A historical anecdote often told in laboratories captures the emotional reality of the change. Early practitioners describe the moment of seeing a vitrified specimen as a kind of relief: the background looked “clean,” the expected objects appeared where they ought to be, and the endless anxiety about whether preparation had turned biology into artefact loosened its grip. This was not the thrill of a single spectacular image, but the quieter satisfaction that the method had become reliable enough to trust. That trust is what allows a community to invest years in improving detectors, optics, and computation. Without a reliable way to present molecules in a near-native state, those later improvements would have been polishing the wrong mirror.

It is tempting to treat vitrification as merely a technical prelude, a practical step that happens before the “real science” begins. But it is better understood as a redefinition of what the electron microscope could be for biology. A light microscope looks at living cells in water without much drama because photons do not require vacuum. An electron microscope, by its nature, seems to demand a trade: resolution in exchange for life. Dubochet’s frozen films did not make the specimen alive, but they preserved an arrangement so close to life that the trade no longer felt like a betrayal. The specimen was no longer a dehydrated residue or a stained silhouette. It was, in a meaningful sense, an arrested aqueous state.

The transformation also changed the role of the grid. The grid stopped being merely a holder and became a stage with strict physical demands. The support film had to be compatible with forming a thin, even aqueous layer. The holes had to be the right size to span films without tearing. The metal had to withstand thermal shock.

The act of blotting became a controlled gesture rather than a casual wipe. And the plunge into cryogen—so fast, so decisive—became the moment when physics and biology shook hands.

A final, often overlooked detail is that vitrification made the water *useful*. Not useful as a solvent, which it had always been, but useful as part of the measurement. In a vitrified sample, water becomes a structural support: an amorphous solid that holds molecules apart from the air and apart from one another, minimizing flattening and aggregation. It is as if the specimen has been embedded in a clear resin that is made of its own environment. The method does not add a foreign scaffold so much as harden the native one.

That is why the breakthrough can be described without exaggeration as a frozen epiphany. The obstacle had been sitting in plain sight—water, the most familiar substance in biology—and the solution was not a new kind of microscope but a new way to ask water to behave. Once water could be persuaded to freeze without turning into crystals, the electron microscope finally had access to the molecules biologists actually cared about, held still in a thin, glassy sheet that was strong enough to survive the vacuum and gentle enough to preserve form. The world inside the droplet, momentarily alive with thermal motion, could be stopped without being shattered.

7.2 The Philosophy of the Faint Signal

A strange etiquette governs electron microscopy of biological matter. The microscope is powerful enough to resolve the spacing between atoms, yet the person at the controls must behave with restraint. The reason is brutally simple: an electron beam does not merely illuminate. It collides, excites, ionizes, and fractures. If the dose is high enough to make a crisp, contrasty picture in a single exposure, it is usually high enough to scramble the very arrangement one hopes to record. Robert Glaeser's early work (1971), and later reviews with colleagues, gave this limitation a firm physical footing: for soft, hydrated specimens, there is a point beyond which "looking harder" becomes an act of destruction.

Low temperature helps. It slows some of the chemical cascades that follow ionization and can delay the onset of obvious damage, which is one reason vitrified specimens are kept so cold. But cold does not repeal physics. At best it buys time and preserves order a little longer. The uncomfortable consequence is that biological electron microscopists are forced into a kind of self-denial: they must keep the electron exposure low enough that the specimen remains interpretable, and accept that each image will be miserably noisy.

To someone meeting the field for the first time, this seems like a dead end. A microscope is supposed to turn the invisible into the visible. Here, the price of not ruining the sample appears to be an image that barely deserves the name. A low-dose micrograph of a protein in vitreous ice can resemble a grey sky speckled with faint shadows. The shadows are not vivid outlines, but slight statistical preferences in where electrons land. If you did not know the experiment, you might not know that anything was present at all.

Joachim Frank's central insight was to treat this apparent failure as a data problem rather than an aesthetic one. A single low-dose image might be hopeless, but many such images—of many copies of the same molecule—could be combined to reveal what no single picture could show. The philosophy was almost democratic: the truth would not come from a heroic exposure or a perfect specimen, but from the collective testimony of a crowd.

That crowd exists because biology is generous in its own way. A purified sample contains enormous numbers of copies of the same protein complex. In a vitrified layer, those copies are embedded in random orientations, like tiny coins suspended in a clear solid, each tilted differently. Each copy offers a projection: a two-dimensional view of a three-dimensional object. Each view is faint, because it has been recorded under low dose. But faint does not mean empty. It means the signal is buried in noise.

Noise, here, is not a metaphor. It has sources that can be named. Some of it comes from the inherently probabilistic nature of electron detection: count a small number of electrons and the fluctuations are large. Some of it comes from the specimen itself: the surrounding vitreous ice contributes its own weak scattering, and the molecules are not perfectly rigid. Some of it comes from the instrument and the recording medium: the way electrons are converted into a measurable image can lose information before any clever algorithm gets a chance.

This last point often surprises readers because it sounds too mundane to matter. Yet detector performance is part of the moral economy of low-dose imaging. If you can only afford a small number of electrons, then you want to waste as few of them as possible. In the language of microscopy, the question becomes: how efficiently does the detector convert incoming electrons into a faithful image across different spatial scales? Researchers such as McMullan and colleagues formalized this in terms like detective quantum efficiency (DQE) and the modulation transfer function (MTF). The underlying idea can be captured without the equations. Imagine photographing a dimly lit street at night with two cameras. Both are pointed at the same scene for the same short exposure. One camera has a sensor that turns nearly every incoming photon into a usable electronic signal; the other has a sensor that turns many photons into heat or blur. The second camera does not just produce a darker picture. It loses fine detail—the texture of brickwork, the edges of street signs—information that cannot be recovered later by sharpening. In low-dose electron microscopy, the "dim street" is the dose-limited specimen, and the "camera" is the detector. A detector that preserves faint, high-frequency information does not merely improve image quality; it changes what kinds of reconstructions are possible from the same fragile sample.

Frank's work took place at a time when neither detectors nor computers were as forgiving as they are now. Much of the early craft involved coaxing meaning from images that, by everyday standards, were dreadful. The key move was to stop asking whether any single particle image looked interpretable and to start asking whether the *set* of particle images contained consistent information.

The first obstacle is that particles are not conveniently aligned. In a micrograph, each molecule is at a different position, rotated differently in the plane of the image, and projected from a different three-dimensional orientation. If you average such images naïvely, you do not reveal structure; you wash it away. The average of many unaligned faces is not a clearer face. It is fog.

Frank helped make alignment and classification the heart of the problem. Alignment means estimating how each particle image should be shifted and rotated so that comparable features overlap. Classification means recognizing that not all particles belong in the same pile: they may represent different views, different conformations, or simply different levels of quality. Once similar images are grouped, averaging becomes meaningful, because it boosts what is shared and suppresses what is random.

None of this is trivial, because the very reason you need alignment and classification is that each image is weak. When the picture is dominated by noise, how can you reliably determine whether one particle is rotated by 17° relative to another? How can you tell whether a faint shadow is a genuine feature or a chance fluctuation? Frank's genius was to insist that the problem could be made mathematically well-posed: the noise was not an enemy to be wished away, but a property to be modeled. If one treats each noisy particle image as a measurement of an underlying signal plus random error, then it becomes possible—through correlation methods, statistical criteria, and iterative refinement—to estimate the alignment parameters that best explain the data. The computations are demanding, but the logic is disarmingly simple: if an alignment is correct, similar images will reinforce one another; if it is wrong, they will not.

There is a human story embedded in that insistence. In the 1970s and early 1980s, the culture of electron microscopy was still strongly visual. A good microscopist was someone who could prepare a specimen and produce an image that looked persuasive to the eye. Frank's approach, developed over years and refined through both theory and software, asked the community to shift its trust from a single compelling picture to an argument built from many unimpressive ones. That is not an easy change. Scientists are trained to be skeptical of what they cannot see, and here they were asked to accept that the real signal might only appear after computation.

A concrete analogy can help. Think of trying to identify a quiet melody in a noisy room. If you listen for ten seconds, you might catch nothing but chatter and clinking glasses. If you record the room for an hour, you still have a noisy recording. But if you suspect that the same melody is being played repeatedly, you can segment the recording into many short clips, align them in time, and average them. Random conversations cancel; the repeated notes add. The melody becomes audible not because any one clip was clean, but because repetition allows you to separate pattern from coincidence.

In cryo-EM, the “melody” is the particle’s projection. The segmentation is particle picking—extracting small boxed images of individual molecules. The alignment is spatial rather than temporal. The averaging is literal averaging of pixel values after alignment. And the repetition is provided by the biochemistry that yields thousands to millions of copies of the same object.

The step from averaged projections to a three-dimensional map has its own conceptual charm. A three-dimensional object is uniquely determined, in principle, by sufficiently many projections taken from different angles—an idea familiar from medical CT scans. The mathematics linking projections to 3D reconstructions is well developed, and by the time single-particle cryo-EM began to mature, there was a clear theoretical footing: the images are projections governed by known physics, and reconstruction can proceed by using their relationships in Fourier space. One does not need to follow the full Fourier machinery to grasp the intuition. If you know what a shadow looks like when an object is lit from different directions, you can infer the object’s shape. The challenge is not that the inference is impossible; it is that the shadows are faint and the object is tiny.

This is where Frank’s crowd-based approach becomes more than a trick. It becomes a way of thinking about what counts as “seeing.” In ordinary photography, a single exposure can be treated as a direct representation. In low-dose cryo-EM, a single exposure is closer to a clue. The representation emerges from a statistical consensus across many clues, constrained by the physics of image formation.

That shift has ethical overtones as well as technical ones. The dose limit is not a mere inconvenience; it is a boundary set by the specimen's vulnerability. One cannot argue with it by bravado. Frank's methods respect the boundary by extracting more information per electron, not by spending more electrons. The sample is treated as something to be listened to gently, repeatedly, until its consistent features become distinguishable from the hiss.

A historical anecdote captures the mood of the early field. Microscopists sometimes described their particle images with a kind of embarrassed honesty: "There's something there, if you squint." The squint, in Frank's view, was not a human squint but a computational one. He and others developed software frameworks that made it possible to process large collections of particle images systematically, replacing the subjective judgment of "this looks good" with quantitative criteria for alignment, variance within a class, and improvement of signal-to-noise through averaging. The result was not only better reconstructions, but a new style of argument. You could show that a feature persisted across independent subsets of data; you could test whether an apparent detail vanished when alignment was perturbed; you could detect heterogeneity by observing that a single class average was unstable and wanted to split into multiple consistent classes. The data themselves could be asked questions.

The faint-signal philosophy also changes the way one interprets disappointment. A bad day in the microscope might yield micrographs that look worse than usual: more drift, thicker ice, lower contrast. In the old visual culture, that could be the end of the story. In the crowd-based culture, one asks a different set of questions. How many particles are usable? Are there enough that the averages will still converge? Is the noise truly random, or is it structured in a way that will mislead classification? A low-quality dataset is not automatically useless; it is a dataset with different statistical limits. This does not remove the need for good experiments, but it widens the range of what can be salvaged into knowledge.

The deeper point is that single-particle cryo-EM, as Frank helped shape it, made peace with the idea that visibility can be emergent. The object is not painted onto the detector in a single decisive moment. Instead, the object is inferred from many weak measurements, each taken under the constraint that the specimen must remain intact long enough to contribute its whisper to the chorus.

When the approach works well, the outcome can look miraculous: a crisp three-dimensional density map emerging from what began as a set of muddy specks. But the miracle is disciplined. It rests on accepting the specimen's fragility, respecting dose limits, insisting on statistical rigor, and building methods that let the consistent information accumulate without being drowned by randomness. The faint signal is not a flaw to be apologized for; it is the price of honesty when the thing you want to observe is easily harmed by observation itself.

7.3 Proof Against Pessimism

There was a time when the most sensible response to biological electron microscopy was a polite shake of the head. The objection was not aesthetic, but physical. Electrons interact strongly with matter. That is why they can, in principle, reveal extremely small features. It is also why they do not behave like a harmless beam of light passing through glass. For biological molecules, strong interaction means two things at once: good contrast is possible, and damage is unavoidable. Glaeser's early dose-limit arguments made the dilemma explicit: to record an image with the clarity one might want, one would need an exposure that would, by the same act, erase the very arrangement being recorded. It was easy to turn that into a gloomy conclusion: perhaps electron microscopy could show outlines, or large shapes, but it would never be a tool for atomic interpretation of proteins.

Richard Henderson's role in the story was not to deny the difficulty, but to insist that it did not amount to impossibility. The distinction matters. Skepticism is healthy when it keeps people honest about what their instruments can and cannot do. Pessimism is something else: a belief that the ceiling is permanently low, so there is little point in reaching. Henderson helped change the ceiling. Not by offering a motivational slogan, but by producing work that forced the community to revise what it thought the physics permitted.

A key part of his strategy was choosing a specimen that behaved unusually well. Many proteins are flexible, heterogeneous, and reluctant to adopt a single, repeated arrangement. The particular membrane protein that became Henderson's benchmark, bacteriorhodopsin, offered something rare: order. It lives naturally in a membrane and, in certain organisms, forms tightly packed arrays in what is often called purple membrane. Those arrays can behave as two-dimensional crystals: not three-dimensional blocks like salt or quartz, but sheets with repeating structure. Repetition is a gift in a dose-limited world. If you can distribute the necessary information across many copies arranged in an orderly pattern, then you can reduce the exposure per copy while still accumulating enough signal overall.

This decision can sound like a technical convenience, but it carried a philosophical edge. If the pessimistic claim was that biology could never be seen at near-atomic detail by electrons, one good counterexample would be enough to puncture it. Henderson's question became: can electron microscopy, under the best circumstances we can engineer, deliver information that supports an atomic model of a protein? If the answer was yes, then the burden of proof shifted. The conversation would no longer be about whether it was possible in principle, but about how widely the success could be extended beyond cooperative specimens.

Bacteriorhodopsin also mattered because it is not an exotic curiosity. It is a light-driven proton pump: a protein that uses light energy to move protons across a membrane, helping the organism harvest energy. Membrane proteins are medically important and experimentally notorious. They are often hard to crystallize in three dimensions, hard to keep stable, hard to image. A method that could handle a membrane protein at high resolution would not merely check a box; it would hint at a broader future.

Henderson and colleagues' 1990 work with Baldwin produced what was widely treated as a decisive moment: an atomic model of bacteriorhodopsin derived from high-resolution electron cryo-microscopy/electron crystallography. The phrase is dense, but the achievement is easy to appreciate. An atomic model is not just a pretty rendering. It is a set of coordinates that explains the data: where the helices lie, how the side chains pack, how the protein's internal cavity is shaped, how the molecule might change when it does its job. To build such a model from electron microscopy required that the images preserve information at spatial frequencies corresponding to individual amino-acid side chains. It required that the specimen remain ordered enough, and stable enough under low-dose conditions, that those faint high-resolution signatures were not washed away.

The deeper point was rhetorical in the best sense: it was an argument made with evidence. Up to then, a critic could say, "All this talk of atomic detail is wishful thinking; the beam ruins the sample before you get there." Henderson's work replied, effectively, "Here is a protein, here are the data, here is a model that accounts for them." The claim was no longer speculative.

To understand why this mattered so much, it helps to picture what counts as "detail" in a protein map. At modest resolution, a membrane protein can be recognized as a bundle of rods, because alpha helices are long, cylindrical features. That alone is useful; it tells you how the protein spans the membrane. But helices are a kind of coarse architecture. To interpret mechanism, you want to know which side chains face a channel, where polar groups cluster, where a retinal molecule binds in bacteriorhodopsin, how the protein might rearrange. Those are features that live closer to the atomic scale. Henderson's accomplishment was to push electron methods across the boundary from "helix bundle as a silhouette" to "helix bundle as a chemically interpretable structure."

This did not happen by stubbornness alone. It required discipline in how images were recorded and interpreted. Low-dose imaging is not just turning down the beam; it is designing the whole experiment around scarcity. You decide how many electrons you can spend, and then you plan how to spend them so that the limited signal you gather is not squandered. You also need to respect the strange ways specimens behave under irradiation at cryogenic temperatures: the gradual loss of high-resolution information, the subtle changes that arrive before obvious bubbling or collapse. The goal becomes to harvest the early, high-frequency information before the sample's internal order degrades.

Order itself was supplied by the two-dimensional crystal. In a single-particle experiment, you rely on averaging across many separate particles that are not physically linked. In a crystal, the averaging is built into the specimen: repeating units reinforce one another. But crystals introduce their own obligations. The lattice must be intact; distortions must be understood; images must be recorded at multiple tilt angles to sample three-dimensional information. Electron crystallography is demanding. Yet it offered a pathway to high resolution that did not require pretending the dose limit did not exist.

There is an often-overlooked psychological detail in how a field changes its mind. A claim of impossibility can survive many partial successes. Each improvement is dismissed as a special case or a fragile demonstration. What changes minds is not a single sharp picture, but a chain of results that shows persistence and scalability: the same specimen improving over time, the same methods being refined, the same physical limits being negotiated rather than denied. Henderson's work on bacteriorhodopsin functioned as that chain. The 1990 atomic model was a landmark, and later high-resolution electron crystallography of bacteriorhodopsin by Kimura and the Fujiyoshi group in 1997 reinforced the message: this was not a one-off miracle. The methods could be pushed, validated, and improved.

A concrete laboratory scene makes the achievement more tangible. Imagine a researcher sitting in a dim control room beside a transmission electron microscope. The specimen, prepared on a grid and cooled to cryogenic temperatures, is inserted into the vacuum. The beam is aligned and blanked, unblanked briefly to search for a promising region, blanked again. Every exposure is a decision. A too-generous exposure yields a pleasing image and a ruined sample. A too-cautious exposure yields an image so noisy it is barely usable. The experimenter is constantly negotiating with invisibility, collecting data that do not reward the eye immediately. Henderson's results told those experimenters that the negotiation was worth it: that the faint patterns contained atomic truth if handled with enough care.

This had consequences beyond bacteriorhodopsin. Once the community accepted that electrons could, under the right conditions, deliver near-atomic information from biological matter, the attention naturally turned to what those "right conditions" might be for less cooperative specimens. The two-dimensional crystal was a special ally, but it also posed a challenge: many important proteins do not form such crystals. If near-atomic interpretation depended on crystal order, perhaps the pessimism would return under a new disguise: possible for a few crystalline membranes, not for the broader molecular world.

The later rise of single-particle cryo-EM to near-atomic resolution is sometimes told as if it rendered electron crystallography obsolete. That misses the continuity. Henderson's bacteriorhodopsin work established something like a proof-of-principle for the entire enterprise: the microscope, the beam, the cold specimen, and careful computation could, together, reach the scale where atoms are not merely decorative. When single-particle work began to surge, aided by improved detectors and computational strategies, it was building on a changed intellectual landscape. The question was no longer "Can electrons ever do this for biology?" but "How do we make this routine for many molecules?"

One might expect that the remaining obstacles were purely mechanical: better microscopes, better cameras, faster computers. Henderson's story shows why that is incomplete. The obstacles were conceptual as well. When the signal is weak, it is easy to fool yourself. A reconstruction can appear to sharpen as you push algorithms harder, yet the sharpening may reflect overfitting—mistaking noise for structure. The vulnerability is not a moral failing; it is built into the mathematics of trying to infer a three-dimensional object from many noisy projections. As methods matured, the field needed ways to make refinement robust, to prevent wishful detail from creeping into maps.

That is part of why algorithmic stabilization mattered. Scheres's RELION framework (2012), with its Bayesian, regularized likelihood approach, became influential not because it made a new kind of picture possible in a magical sense, but because it helped make high-resolution reconstruction more disciplined under low signal-to-noise conditions. Regularization is, in everyday terms, a way of penalizing overly complicated explanations when the data are uncertain. It embodies a kind of humility that matches the dose limit: if the data cannot support a sharp feature, the algorithm should not invent it. This ethos harmonizes with Henderson's broader contribution: ambition tethered to physical reality.

There is also a quieter lesson in Henderson's persistence with a single benchmark specimen. In science, it is fashionable to chase novelty. Benchmarking can look repetitive. Yet benchmarks are how a community learns what "better" means. Bacteriorhodopsin became a kind of measuring stick for electron methods: a sample whose behaviour could be studied, whose order could be exploited, and whose reconstructions could be compared across years. The work turned pessimism into a technical agenda. If something blocked higher resolution, one could ask what it was: specimen disorder, beam-induced changes, recording inefficiency, computational misalignment, overfitting. The problems became separable. And once problems are separable, they can be solved.

In the end, Henderson's proof was not simply that a protein structure could be obtained. It was that the dose-limited, low-contrast world of biological electron imaging still contained a path to atomic interpretation, provided one chose specimens wisely, measured carefully, and refused to accept that faintness and fragility implied futility. The photograph might never be bold in the way a stained micrograph is bold, but boldness was never the point. The point was whether the data, extracted gently from a vulnerable molecule, could justify the placement of atoms and the telling of a chemical story.

That possibility changes the emotional texture of the field. A noisy image is no longer an embarrassment. It is the beginning of an argument, one that can culminate in a model detailed enough to explain how light moves protons, how a channel opens, how a drug might bind. Pessimism does not vanish entirely—it rarely should—but it loses its authority when a single protein, held in an orderly membrane sheet, can compel the conclusion that the atomic scale is not off-limits to electrons.

7.4 Stockholm Names the Revolution

On an October morning in 2017, the world's attention briefly swung toward a familiar stage: Stockholm, the Nobel announcements, the careful phrasing that turns years of work into a few sentences fit for radio and headlines. The Nobel Prize in Chemistry that year went to Jacques Dubochet, Joachim Frank, and Richard Henderson, explicitly for developing cryo-electron microscopy into a method for high-resolution structure determination. The prize did something larger than congratulate three careers. It told a wider scientific public—chemists, biologists, physicians, funders, students deciding what to learn—that cryo-EM had crossed a threshold. It was no longer an ingenious specialty living at the edge of microscopy. It had become a central way of turning molecules into knowledge.

The Nobel committee's choices are always read as a story, even when the committee insists it is merely recognizing achievement. Here the story was unusually tidy: three names corresponding to three bottlenecks. Dubochet made water behave by vitrifying it. Frank made faint images trustworthy by treating them as members of a statistical crowd. Henderson supplied proof that near-atomic interpretation was not a fantasy. Those contributions were not independent strokes of genius that happened to share a label. They were complementary answers to a single problem: how to look at biological molecules in a state close to native, without destroying them, and still extract enough information to build a three-dimensional account.

It is tempting to treat "revolution" as a word applied after the fact, a flourish added by committees and press offices. Yet revolutions in science are not only changes in results; they are changes in expectations. Before cryo-EM became reliable at high resolution, a biochemist might build an entire research program assuming that certain kinds of proteins—large, flexible complexes; membrane proteins that refuse to crystallize; fleeting assemblies that fall apart under purification—would remain structurally opaque. The standard techniques, such as X-ray crystallography and NMR spectroscopy, are powerful but selective in what they reward. Crystallography, in particular, is a superb method when crystals cooperate and a stubborn barrier when they do not. Cryo-EM changed that selective pressure. Suddenly, "too hard to see" stopped being a description of nature and became a description of the current state of technique, which might improve next year.

The Nobel announcement mattered because it amplified that shift beyond the community that had lived through the struggle. Within the field, the changes were already evident: laboratories upgrading microscopes, computational pipelines becoming standardized, graduate students choosing cryo-EM as a default tool rather than a gamble. But outside, many still thought of electron microscopy as a beautiful way to image cells, or a source of striking pictures of viruses, not as a general method for making atomic models. A Nobel Prize operates as a kind of translation device. It takes a technical claim—"these reconstructions reach near-atomic resolution under dose constraints using vitrified specimens"—and recasts it as a cultural fact: "this is now a foundational technique."

There is a human dimension to this translation. For decades, cryo-EM demanded a temperament that was not universally rewarded: patience with fragile samples, tolerance for images that looked unimpressive, faith that computation could turn faint traces into something precise. The work often unfolded in the shadows of more established disciplines. Crystallographers could point to elegant diffraction patterns and well-validated models. Electron microscopists often had to explain, repeatedly, why their results were not artefacts of staining, dehydration, or wishful reconstruction. The Nobel did not erase those struggles, but it changed how they were remembered. What had seemed like a niche persistence began to look, in retrospect, like the slow building of a bridge that everyone now wanted to cross.

A concrete example shows why the prize resonated beyond microscopy circles. Consider how quickly molecular structures can become public events when they matter for health. A viral protein structure, once a specialized object studied by a few labs, can become the basis for vaccine design, drug screening, and public discussion about variants. Cryo-EM is unusually suited to such urgent moments because it can handle large complexes and membrane proteins that are difficult to crystallize, and it can deliver three-dimensional maps without requiring that molecules form large, perfect crystals. When the Nobel committee described cryo-EM as a revolution, it was also endorsing a new tempo of discovery: structures that once took years of trial and error could, for some targets, be obtained on a timescale that matches biological urgency.

The Nobel Foundation's background materials, which accompany the announcement, play a particular role in shaping memory. They do not merely list papers and dates; they curate a lineage. In this curated lineage, cryo-EM's success is not attributed to a single gadget or a single heroic laboratory. It is framed as a system: specimen preparation, imaging under dose constraints, computation to reconstruct three-dimensional information, and validation to make sure the "detail" is real. That framing is more than diplomatic. It is accurate to the way the field actually advanced. Improvements came not only from sharper lenses but from better detectors, more stable stages, more sophisticated algorithms, and above all, a shared discipline about what could be claimed from noisy data.

The press reaction in 2017, captured in contemporaneous science journalism, had its own emphases. Reporters needed examples and metaphors that could carry the idea to readers who might not know a ribosome from a receptor. Viruses were popular protagonists: they are visually compelling and socially relevant, and the idea of "seeing" a virus at near-atomic detail has a natural drama. Molecular machines inside the cell offered another kind of hook: a way to talk about life as engineered, without implying design. In many of those accounts, cryo-EM was portrayed as a camera that suddenly became sharper. That is an understandable simplification, but it risks obscuring the true nature of the change. The revolution was not just improved optics; it was the hard-won realization that you could never simply blast the specimen with electrons until it yielded a clear photograph. You had to accept the faintness, respect the damage limits, and build your picture out of many careful, partial views.

One historical anecdote from the community, often retold with affectionate exasperation, concerns the way early cryo-EM practitioners defended their work to skeptics. They would show a low-dose micrograph and be met with a skeptical stare: "Where is the molecule?" The answer was that the molecule was there, but it was distributed across thousands of such images, and its clarity would emerge only after alignment and averaging. For a long time, that answer sounded to outsiders like a plea for faith. The Nobel announcement signaled that the faith had been converted into a mature methodology: not a promise that computation could do magic, but a demonstration that it could extract consistent information from the constraints imposed by physics.

The Nobel also changed institutional behaviour in ways that are easy to underestimate. Universities and research institutes respond to prestige with investment. A high-end cryo-electron microscope is expensive not only to purchase but to house, maintain, and staff. It demands vibration control, electromagnetic shielding, stable power, liquid nitrogen logistics, and specialist expertise. Before cryo-EM was widely seen as a mainstream structural method, such investments were hard to justify unless an institution already had a strong microscopy culture. After 2017, the calculus shifted. Cryo-EM facilities became emblematic of modern molecular research in the way synchrotron access had once been. New centres were built; existing ones expanded; training programs grew. The Nobel did not cause the technical advances, but it accelerated their diffusion by making cryo-EM an obvious bet for administrators who might otherwise hesitate.

There is also a subtler effect on the self-conception of researchers. A method's prestige influences who enters the field. When cryo-EM was regarded as a specialist craft, it attracted people who enjoyed instrumentation, who could tolerate tinkering, who found satisfaction in coaxing order from stubborn materials. As it became more central, it also attracted a different mix: biologists with specific molecules they wanted to understand, computational scientists drawn by the data challenges, chemists interested in mechanism and drug design. This broadening can enrich a field, but it also changes its internal culture. The older generation had learned to distrust beautiful pictures and to demand rigorous validation. Newcomers, excited by the apparent ease of producing models, needed to be taught the discipline that the pioneers had learned the hard way: that noise can be persuasive, that over-interpretation is a constant temptation, and that resolution numbers are not moral endorsements.

In a sense, the Nobel prize functioned as a naming ceremony. "Revolution" became the common label for a shift that had already happened in practice and was now being acknowledged publicly. Naming matters because it creates a shared reference point. It tells grant panels what counts as contemporary. It tells students what is worth learning. It tells neighbouring fields that a tool is ready to be borrowed. It also shapes the stories scientists tell about their own work. Once cryo-EM was officially revolutionary, researchers could frame new results not as isolated technical feats but as part of a larger transformation: a new way of making the invisible accountable.

The tripartite framing of the prize also quietly protected the field from a common misunderstanding: the belief that a single advance, such as a better detector or a faster computer, is the whole explanation. By honouring specimen vitrification, computational reconstruction, and the demonstration of atomic feasibility, the Nobel committee placed cryo-EM's success on three legs. Remove any one and the stool wobbles. Without vitrification, the specimen would be compromised before imaging. Without computation, the faint images would remain faint, their information inaccessible. Without proof that near-atomic interpretation was possible, the method might have remained a way to obtain coarse shapes, valuable but not transformative. The prize argued, in effect, that modern cryo-EM is an ecosystem of ideas and practices, not a single clever trick.

That ecosystem continues to influence how molecular science is imagined. In earlier eras, the dream of "seeing" a protein at atomic detail often came with a tacit assumption: the molecule would need to be coaxed into a crystal, a highly ordered state that is as much an achievement of chemistry as it is a gift from nature. Cryo-EM relaxed that assumption. It allowed people to imagine structures of complexes that are heterogeneous, transient, or difficult to crystallize, and to imagine those structures appearing in multiple functional states rather than as a single frozen pose. The imagination is not mere fantasy; it guides what problems scientists attempt, what collaborations they build, and what kinds of therapeutic strategies they consider plausible.

The Nobel prize in 2017 did not end the challenges of cryo-EM. Sample preparation remains capricious; preferred orientations and ice thickness still frustrate; the dance between signal and damage still governs what can be claimed. But the announcement changed the default mood. It licensed ambition. It told a broad community that the faint signal, the glassy water, and the careful proof of atomic detail were not fragile exceptions. Together, they had become a method sturdy enough to carry the weight of modern molecular questions, from the machinery of cells to the urgent proteins of disease.

Chapter 8

Machines That Translate, Fold, and Fight

The most revealing molecular portraits are often the busiest ones: machines caught mid-task, assemblies too unruly to crystallize, and pathogens whose surfaces hide both vulnerability and threat. Here the reader discovers how cryo-EM turned motion into evidence, how structural biologists learned to combine multiple ways of seeing, and how shared archives transformed private datasets into a common scientific memory.

8.1 The Ribosome Caught in the Act

A ribosome is not subtle. It is one of the largest and busiest molecular machines in any cell, a two-part factory built from RNA and protein, clamping onto a messenger RNA (mRNA) and turning its sequence into a growing chain of amino acids. Cells keep thousands of ribosomes running at once, because life depends on a constant supply of fresh proteins: enzymes, receptors, scaffolds, signals. If the ribosome were a human-made device, it would be industrial equipment, not a delicate instrument.

And yet the first emotional shock, for many people who work with cryo-EM, arrives precisely here: a machine this massive is also this restless. The ribosome does not simply open, read, and shut like a stamp. It flexes, swivels, and steps forward in coordinated motions. It binds helpers, releases them, and briefly holds configurations that exist only to be left behind. In translation (the process of building a protein), the ribosome has to do something that is mechanically non-trivial: keep the mRNA and the transfer RNAs (tRNAs) aligned by three-letter codons while it repeatedly advances by exactly one codon, without slipping, and while it chemically stitches amino acids together into a chain.

For a long time, structural biology tended to reward stillness. A crystal is the ultimate request for order: molecules must agree to repeat in the same pose, over and over, until a macroscopic lattice forms. That demand has produced many of biology's most precise atomic models, including landmark X-ray crystal structures of the ribosomal subunits around the year 2000 by Ada Yonath, Thomas Steitz, and Venkatraman Ramakrishnan. Those structures transformed biology, not least by showing that the ribosome's catalytic core is RNA, and by revealing how antibiotics exploit ribosomal vulnerabilities. But a crystal does not naturally preserve the bustle of translation. Even when an intermediate can be coaxed into a lattice, the very act of crystallizing selects certain conformations and discourages others. The ribosome in motion becomes, by necessity, the ribosome in a particular compromise.

Cryo-EM changed what counted as a compromise. The shift was not simply that ribosomes could be imaged without crystals. The deeper change was that *mixtures stopped being a reason to give up.* A tube of ribosomes translating is not one object repeated thousands of times. It is a population: some ribosomes have a tRNA in the A site (the entry site), some have shifted into the P site (the peptidyl site), some are holding an empty tRNA in the E site (the exit site), some have a helper factor bound, some have just let one go. In older thinking, this was a nuisance called heterogeneity: the sample was "heterogeneous" because it refused to settle into a single uniform arrangement. In cryo-EM, that heterogeneity can become the point. If the molecules occupy multiple states because they are doing their job, then a microscope that can separate states is not merely taking a portrait; it is taking attendance.

The key idea is simple to state. A cryo-EM dataset contains hundreds of thousands to millions of particle images, each one a noisy, two-dimensional projection of a single ribosome trapped in glass ice in whatever conformation it happened to occupy at the moment of freezing. Those particles are not identical, and they are not oriented the same way. The trick is to sort them into groups that *are* internally consistent. The sorting does not happen by hand, and it does not require knowing in advance what states exist. It happens through statistical pattern-finding that asks which images resemble each other.

The first sorting step is often called *2D classification*. Images are grouped by similarity into classes, and each class is averaged to produce a clearer representative view. This already does two valuable things. It throws away obvious junk (ice contamination, broken particles, images where alignment failed). And it shows whether the dataset contains more than one "kind" of ribosome view: perhaps some particles carry extra density from a bound factor, perhaps some have a rotated subunit orientation, perhaps some are missing a subunit entirely. Even for a non-specialist, it is worth pausing to appreciate what the computer is being asked to do. Each particle image is a dim shadow of a molecule a few tens of nanometres across, recorded at a dose low enough to avoid destroying it. And yet, by aligning and averaging many such shadows, the ribosome's outline, cavities, and protruding stalks begin to appear with a solidity that feels almost undeserved.

Then comes the step that turns sorting into mechanism: *3D reconstruction* combined with *3D classification.* A reconstruction algorithm uses many 2D projections from many orientations to infer a 3D density map: a volumetric estimate of where the molecule's mass sits. But if the dataset contains multiple conformations, a single reconstruction will blur them together, smearing moving parts into fog. 3D classification confronts that directly by asking for multiple reconstructions at once, each one built from a subset of particles that share a conformation. Instead of forcing the ribosome to be one thing, the analysis allows it to be a family of related things.

This is why the ribosome became such a persuasive demonstration of cryo-EM's distinctive power. Translation is a cycle with recognizable landmarks, and the ribosome is large enough to anchor alignment well. A moving tRNA or a rotated subunit is not a subtle side-chain flip; it is a substantial rearrangement. What once counted as annoying variability becomes interpretable variation: the ribosome is not failing to behave; it is proceeding through its working states.

It also helps that the ribosome offers an intuitive cast of characters. The mRNA is the tape of instructions. Each tRNA is an adaptor that matches a codon on the mRNA to a specific amino acid, and the ribosome is the stage on which those adaptors arrive, pair, react, and depart. The ribosome's two subunits wrap around the mRNA and the tRNAs; bacteria have a small subunit and a large one that together form the 70S ribosome. In translation elongation, a charged tRNA enters the A site, the ribosome joins its amino acid to the growing peptide chain held by the tRNA in the P site, and then everything shifts: the tRNAs move from A to P and from P to E, and the mRNA advances by one codon so the next instruction is ready.

The shift is called *translocation*, and it is where the ribosome's restlessness becomes unavoidable. In bacteria, a protein called EF-G (elongation factor G) binds to the ribosome and helps drive this step, powered by GTP. Biochemists could infer for decades that translocation involved intermediate states: partial movements, hybrid configurations, steps with distinct timing. Single-molecule measurements could even watch transitions in real time as changes in fluorescence, like a heartbeat traced on a monitor. But it is one thing to measure that a transition exists and another to see the molecular geometry of the transition.

One of cryo-EM's great gifts to ribosome biology has been to put shapes to these inferred intermediates. Cryo-EM studies in the early 2010s captured ribosomes with tRNAs in *hybrid states*—configurations where parts of a tRNA have shifted relative to one subunit while other parts still contact the other subunit, as if the molecule is midway through a step and temporarily straddling positions. This is not an artistic metaphor; it is a literal description of a molecule whose contacts cannot be described as purely "A site" or "P site" in the traditional tidy diagram. The ribosome itself participates by undergoing intersubunit rotation, often called *ratcheting*: the small subunit rotates relative to the large subunit, changing the geometry of binding pockets and making certain moves easier. Multiple moving elements—tRNAs, subunits, flexible ribosomal RNA segments, and factor proteins—work as a coupled system. A ribosome is not a rigid clamp; it is a machine whose parts are designed to move together.

How do you catch a fast, multi-step process that exists to keep moving? One strategy is to make the ribosome pause. The pause can be induced with antibiotics, with mutations, or with nucleotide analogs that mimic GTP states and alter EF-G's timing. This is where structural biology and medicine have long been intertwined in a practical way: antibiotics are not only therapeutics, they are also tools for trapping machinery. Some drugs effectively jam the ribosome at a particular stage, stabilizing an intermediate that would otherwise be fleeting. What is ethically complicated in the clinic (a molecule that disrupts a vital cellular process) is experimentally useful at the bench (a molecule that stops the machine mid-action so it can be examined).

Another strategy is more daring: rather than trapping a single state, accept that the tube will contain many states and use computation to separate them. This approach aligns with how translation actually proceeds. In a population of ribosomes, not all molecules step in synchrony. If the reaction is allowed to proceed for some time and then frozen, the sample is a cross-section of the cycle: some ribosomes are early, some late, some waiting for a factor, some just released one. In older methods, that asynchrony was an obstacle. In cryo-EM it can be a source of mechanistic information, provided the classification is reliable.

Reliability is not a given; it is earned by careful handling of the data and by scepticism about what classification can and cannot guarantee. A computer will always find classes, because it is asked to. The hard question is whether those classes correspond to meaningful physical states rather than artefacts of noise, misalignment, or overinterpretation. That is why ribosome work has been such a proving ground: the field could cross-check structural classes against decades of biochemical knowledge, antibiotic effects, and kinetic measurements. If a class shows EF-G bound and a rotated subunit, and that matches what other experiments predict for a translocation intermediate, confidence rises. If a proposed class contradicts established constraints, it demands caution.

As detectors and analysis methods improved, another possibility became practical: *time-resolved* cryo-EM, where the experiment is set up to take snapshots at defined reaction times. A translation reaction can be started by mixing components, allowed to proceed for a chosen "aging" period, and then rapidly frozen. By repeating this at different times and then classifying the resulting particles, researchers can enrich early intermediates in one dataset and later intermediates in another. Time-resolved cryo-EM has used this approach to visualize ribosomal translocation with EF-G and GTP, recovering multiple intermediates spanning pre-translocation, EF-G-bound intermediates, and post-translocation states. The achievement was not merely technical. It offered a bridge between structure and kinetics: the population of particles in each class changes with time in a way that can be compared to biochemical rates. Structure ceases to be only an endpoint and becomes a language for describing progression.

To understand why this feels like a kind of cinema, it helps to be precise about what is, and is not, being claimed. Cryo-EM does not film a single ribosome moving through time. Each particle image is a different molecule, frozen at a different moment. The "movie" is assembled statistically: a sequence of frames reconstructed from ensembles. If you want to be strict, it is not a movie of one actor but a montage assembled from many performances. Yet the ribosome is such a stereotyped machine, and its states recur so reliably, that the montage can be compelling. When classification yields a set of structures that differ by systematic movements—a tRNA shifted, a subunit rotated, a factor attached or released—the mind stitches them into a coherent action.

There is also a conceptual payoff that is easy to miss. For years, a "structure" in biology was often treated as a noun: the structure of the ribosome, the structure of a channel, the structure of a virus. The singular noun invites the idea of a canonical arrangement. The ribosome resists this, and cryo-EM makes the resistance visible. What you get is not one structure but a *structural pathway*: an itinerary through a landscape of shapes. The pathway is not arbitrary; it is constrained by chemistry and geometry. The ribosome cannot put a tRNA anywhere it likes, because the codon-anticodon pairing must be maintained and the catalytic centre must remain positioned to form peptide bonds. Motion is guided, not random. But it is still motion, and the correct unit of description becomes something like "the ensemble of states used during translocation" rather than "the ribosome's structure."

That shift in language matters, because it changes the kinds of questions a biologist can ask. If you are stuck with one pose, you ask where the active site is and what residues line it. If you have multiple poses, you ask what moves, when it moves, what controls the branching between alternatives, and where a small molecule might intervene to freeze the machine in a non-productive state. You begin to think about translation the way an engineer thinks about a mechanism: not only the parts list, but the cycle.

The ribosome also offers a humbling lesson about scale. A ribosome is huge by molecular standards, and yet its critical tasks depend on tiny alignments: three nucleotides at a time on the mRNA, a precisely positioned amino acid, a catalytic centre arranged to favour one reaction and discourage others. The machine achieves this by being both robust and flexible. Large movements reposition entire domains; small adjustments in RNA and protein allow correct pairing and chemistry. Cryo-EM, when it separates states rather than averaging them away, makes this dual nature tangible. A ribosome can be seen as a solid object and as a set of coordinated deformations.

There is a final reason ribosome work has an emotional pull, beyond its technical elegance. Translation is the moment where an abstract code becomes physical matter. The genome can seem remote: a text stored in a nucleus or a nucleoid, copied and repaired, sometimes expressed and sometimes silent. The ribosome is where a message becomes a molecule you can weigh, fold, degrade, or misfold. Many diseases begin as errors in this chain of events: a wrong amino acid incorporated, a stalled ribosome, a truncated protein, a misfolded product that aggregates. Many antibiotics save lives by targeting bacterial ribosomes without stopping ours. It is hard not to feel a sense of intimacy when a method lets you see the machinery that sits at this border between information and substance, caught mid-step with its tRNAs in transit and its subunits poised for the next click forward.

8.2 When One Lens Is Not Enough

A molecule can be perfectly real and still refuse to give a single, definitive portrait. Some proteins behave like well-made tools: stable, compact, repeating the same fold whenever you look. Others behave like hands: they open and close, they grip different partners, they change shape as part of their job. The central difficulty is not that scientists lack ingenuity, but that every method for "seeing" molecules is also a way of asking them to hold still, to line up, or to speak through an indirect signal. Different techniques ask different favours of the sample, and the sample answers differently depending on its size, flexibility, and temperament.

Cryo-EM, X-ray crystallography, and nuclear magnetic resonance (NMR) spectroscopy are sometimes presented as rival claimants to the molecular world, as if one will win and the others will fade. In practice, the three techniques resemble different lenses in a camera bag: each has a characteristic view, each introduces its own distortions, and the most reliable picture often comes from using more than one.

X-ray crystallography is the oldest of the three as a routine workhorse, and it still produces many of the most detailed atomic models in biology. The essential bargain is clear. If a molecule can be coaxed into a crystal, X-rays scattered through that orderly lattice can be converted into an electron-density map, and from that map an atomic model can be built with exquisitely precise geometry. At its best, crystallography does not just place a protein's backbone; it resolves alternative side-chain positions, water molecules tucked into crevices, ions perched at active sites. It can tell a chemist why one substitution on a drug strengthens binding and another ruins it.

The cost is that crystals demand agreement. Every molecule in the lattice must occupy the same arrangement, or at least a closely related set of arrangements that can be averaged into a coherent diffraction pattern. Some proteins crystallize readily, as if they were made for it; others refuse for years. Laboratory folklore is full of stories of researchers who grew crystals only after trying dozens of constructs, clipping off a floppy tail, mutating a surface residue, changing salt, changing temperature, adding a ligand, adding an antibody fragment, adding a partner protein, or simply waiting long enough for luck to strike. The joke that crystallography is "applied patience" persists because it carries a grain of truth.

That patience paid off spectacularly in some of biology's most famous moments. When the first high-impact crystal structures of ribosomal subunits appeared around 2000, they revealed the architecture of a machine that had dominated textbook diagrams for decades. But even the success carries its own lesson: the ribosome had to be prepared in ways that encouraged uniformity, and crystals captured particular functional or near-functional arrangements. The method did what it does best, producing a detailed, compelling model. The moving ribosome, the one that actually conducts translation in living cells, remains a shifting population.

Cryo-EM accepts that population more easily. Its sample requirement is not a crystal lattice but a field of isolated particles suspended in vitreous ice. That shift changes the kinds of molecules that can be studied. Large complexes, multi-protein assemblies, and machines with multiple functional states are often natural candidates, because their size makes them easier to align and their heterogeneity can be separated computationally. Cryo-EM can also succeed where crystallography struggles for a mundane reason: the conditions that keep a complex intact in solution are often incompatible with the conditions that promote crystallization. A complex that dissociates at modest concentration, or a membrane protein that requires detergent or lipids to remain folded, may never settle into a crystal. Freezing the intact particles in a thin layer of glassy water can be a gentler request than building a lattice.

NMR, by contrast, does not ask the molecule to freeze or crystallize at all. It asks the molecule to sit in solution and reveal itself through magnetism. Certain atomic nuclei, such as hydrogen-1, behave like tiny magnets. In a strong magnetic field, they occupy discrete energy states, and when nudged with radiofrequency pulses they emit signals that depend on their local chemical environment. By analysing a carefully designed set of NMR experiments, researchers can infer distances and angles between atoms, and from those constraints build structures. Kurt Wüthrich's work in the 1980s helped establish solution NMR as a practical way to determine protein structures, a development that opened a new window onto molecules that simply do not like to crystallize.

NMR's special talent is motion. Because the sample is in solution and the signals can report on dynamics over a wide range of timescales, NMR can detect when a protein is not one structure but a fluctuating ensemble. It can reveal flexible loops, transient binding, and conformational exchange. It can also map local environments: which residues are perturbed by ligand binding, which parts of a protein become more rigid upon complex formation, which atoms experience changing magnetic neighbourhoods as the molecule breathes. There are trade-offs. NMR generally favours smaller proteins and domains, and as the molecule grows, the spectra become crowded and signals broaden. Clever labelling strategies and methodological advances extend the range, but there remains a practical ceiling that makes the technique less suited to massive machines.

These differences can sound like an argument about winners and losers, but a better way to think about it is to ask what each method *produces*, and what it *cannot help producing*. Crystallography produces electron-density maps whose sharpness depends on crystal order and data quality; cryo-EM produces density maps whose clarity depends on alignment, classification, and the number of particles supporting a given state; NMR produces a network of constraints that can lead to a single structure or an ensemble of compatible structures, along with measurements that report directly on motion.

None of these outputs is a photograph. Each is a reconstruction, and each contains uncertainty. In crystallography, a map is shaped by how well the model explains the diffraction data, and refinement can hide a danger: it is possible to "overfit," building features that serve the model more than the data. Crystallography developed sophisticated validation culture partly because its successes were so influential; a beautiful model is persuasive, and persuasion is not the same thing as correctness. In cryo-EM, uncertainty is tied to noise, preferred orientations of particles, and the way classification divides the data. A map can contain regions of differing quality: a stable core with crisp density and flexible arms that blur. In NMR, uncertainty is built into the method's strength: if a protein samples multiple conformations in solution, the constraints may reflect an average, and the most honest description may be a family of structures rather than a single frozen pose.

A concrete way to see the complementarity is to imagine a complicated machine made of both rigid parts and loose joints. Cryo-EM might provide the overall body plan: the arrangement of subunits, the placement of a ligand-bound partner, the global conformational change that defines an "open" versus a "closed" state. Crystallography might then provide the detailed geometry of a specific domain: the exact side chains forming a catalytic pocket, the hydrogen-bond network that positions a critical residue, a bound inhibitor in a precise orientation. NMR might report which loops remain mobile in solution, or which residues experience exchange between conformations, or whether binding induces rigidity in a region that is fuzzy in the cryo-EM map.

Real research often follows this blended logic, sometimes by design and sometimes out of necessity. A lab may solve the crystal structure of a protein domain first, because the domain is small and tractable, and then use that atomic model as a building block when interpreting a cryo-EM map of the full complex. This is not cheating; it is an explicit acknowledgement that different data carry different types of information. If a cryo-EM map has moderate resolution in a region, it may not uniquely determine every side-chain placement. A crystallographic model of that region can provide a plausible, chemically sensible structure that fits the density. Conversely, cryo-EM may reveal that the domain sits in a different orientation in the intact assembly than it ever did in the crystal, warning against naïvely transferring conclusions from a fragment to the full machine.

The blending has become systematic enough to have a name: *integrative* or *hybrid* approaches. The premise is straightforward: many molecular assemblies are too large, too flexible, or too heterogeneous for any one technique to describe completely. Instead of hoping for a single miracle dataset, researchers collect partial constraints and combine them into models that satisfy the full set as well as possible. The data can be of many kinds: cryo-EM density maps, crystallographic structures of components, NMR restraints for domains, small-angle X-ray scattering (SAXS) for overall shape in solution, or biochemical constraints such as known contacts or mutational effects.

One of the most influential additions to this toolkit has been cross-linking mass spectrometry (XL-MS). The concept sounds almost too practical to be profound. A cross-linker is a chemical reagent that can covalently connect two amino acid side chains that come within a certain distance of each other. If a complex is treated with a cross-linker, then digested into peptides and analysed by mass spectrometry, the cross-linked peptide pairs can be identified. Each cross-link implies a spatial restraint: those two residues must be close enough, in the real complex, for the cross-linker to bridge them. XL-MS is a high-impact integrative method because it supplies experimentally derived distance restraints that help place subunits or domains in large, flexible complexes. It will not, on its own, yield a full structure. What it offers is a set of "not too far apart" constraints, scattered across the assembly, that can rule out many wrong arrangements.

Now imagine combining XL-MS with cryo-EM. The cryo-EM map may define an overall envelope for a complex, but some parts could be ambiguous: two similar-sized domains could be swapped, or a flexible arm could occupy multiple positions. Cross-links can break the tie. If residue A on subunit 1 cross-links to residue B on subunit 3, then any model that places those residues on opposite sides of the complex becomes untenable. Cross-links can also validate. A model that fits the cryo-EM density beautifully but violates many cross-links is a warning sign that the map has been over-interpreted or the model has been forced.

The same logic applies to NMR. Suppose a cryo-EM map suggests where a small domain sits, but the domain's internal features are too blurred to model confidently. NMR can provide a high-quality structure of that domain in solution, including information about which parts are mobile. That structure can then be docked into the cryo-EM density, and if the NMR data show that one face of the domain is involved in binding (through chemical shift perturbations, for example), it can guide which orientation is plausible.

This is where modern modelling frameworks become essential. Integrative structural biology is not a casual collage of results; it is a computational discipline that tries to be explicit about what is known, what is assumed, and what remains ambiguous. The Integrative Modeling Platform (IMP) operationalizes integrative modeling as an explicit workflow: define representations, convert data to spatial restraints, sample models, and validate ensembles against data. Several phrases in that description carry philosophical weight.

To "define representations" means deciding how detailed the model should be. One region might be represented atom-by-atom because crystallography has already resolved it. Another region might be represented as a coarse bead model because the data do not justify atomic detail. To "convert data to spatial restraints" means turning measurements into computable constraints: cross-links become maximum distances, EM maps become density-fitting terms, known stoichiometries become constraints on copy number. "Sampling" acknowledges that there may be many models compatible with the data; the goal is not always a single answer but an ensemble of acceptable solutions. Validation, finally, is the bridge between an appealing model and a trustworthy one: does the model satisfy the data it claims to use, and does it predict independent observations?

The emphasis on ensembles deserves special attention, because it marks a cultural change. Older habits of thought pushed toward a single published structure, a single set of coordinates. Yet many complexes do not have one arrangement. They have families of related arrangements, or alternative conformations that interconvert, or flexible regions that cannot be pinned down to one position. Integrative approaches are often more honest about this, because the data themselves have varied resolution and varied specificity. A cross-link does not tell you the angle between two helices; it tells you they are near. A low-resolution map does not uniquely determine side chains; it constrains overall shape. An ensemble model can capture this uneven certainty: some parts tightly defined, others deliberately fuzzy.

This honesty has practical consequences. Drug discovery, for example, often hinges on whether a binding pocket is pre-formed or transient, and whether a protein's surface presents a vulnerable groove only in a particular conformation. A single model can give a false sense of rigidity. An ensemble can show that a "pocket" is really a breathing space that appears and disappears, making some strategies plausible and others misguided.

A historical anecdote captures the human side of these trade-offs. For decades, researchers who loved NMR would speak of the virtue of seeing proteins in solution, "as they are in life," and crystallographers would respond that the rigour and clarity of diffraction were unmatched. The debate could be tribal. Yet as methods matured, the most convincing papers increasingly refused to pick sides. A complex might be described by a cryo-EM map, its subunits refined by crystallography, and its moving parts interrogated by NMR. What once sounded like methodological competition began to resemble good craftsmanship: use the tool that speaks most clearly to the question, and then cross-check with another tool that speaks in a different dialect.

The value of cross-checking becomes clearest when claims are tempting. A density map invites interpretation. If a small molecule is present during preparation, one naturally wants to see it bound. In crystallography, a ligand can often be resolved as a distinct shape in the density, with clear orientation. In cryo-EM, ligands can be harder: occupancy may be partial, the surrounding protein may be flexible, and the local resolution may not support a confident placement. A sensible response is not despair but triangulation: use biochemical assays to confirm binding, use mutagenesis to test whether predicted contacts matter, use alternative structural measurements if possible. The structure is a hypothesis generator and a constraint, not a verdict.

The modern field has also developed a way to make these multi-method stories more than persuasive narratives: it has begun to treat the supporting data and modelling provenance as first-class research objects. Infrastructure for integrative models has matured: PDB-Dev evolved into PDB-IHM with deposition, curation, and validation goals aligned with FAIR principles, and with expectations to deposit associated restraints and modelling provenance. That last phrase—provenance—is quietly radical. It asks not only "what is the structure?" but "how was it obtained, what data constrain it, what alternatives were considered, and how should another researcher reproduce or challenge it?" When multiple methods contribute, transparency is the only way to prevent a hybrid model from becoming an uncheckable authority.

At a dinner party, the simplest way to describe the situation is to borrow a common experience. If you are trying to understand a complex object—a car engine, a musical instrument, a city neighbourhood—one type of image is never enough. A close-up photograph can show you a bolt head but not the wiring harness; an aerial view can show layout but not wear on a belt; a video can show motion but not internal detail. Molecules pose the same problem, only with the added constraint that the "images" must be inferred from delicate signals and careful modelling. The wise response is not to argue about which view is real, but to ask which view answers which question, and which combination of views keeps you honest.

A ribosome caught mid-translocation, a receptor flexing as it couples to its partner, a complex whose subunits associate and dissociate: these are not failures of experimental control. They are the behaviours that make biology work. Seeing them requires a willingness to accept that clarity can be assembled from partial glimpses, and that the sharpest picture is often the one that admits its own remaining blur.

8.3 The Internet of Structures

There was a time when a molecular structure was, in a very literal sense, a picture in a paper. You could admire it, perhaps trace it by hand, and if you were determined you might rebuild it from printed tables of numbers. But the structure itself was not a shared object in the way a gene sequence or a telescope image is shared. It lived inside a journal article as an illustration and a claim: persuasive, often beautiful, and surprisingly difficult for anyone else to pick up and use.

That situation began to change when the field adopted a deceptively simple habit: deposit the underlying data in public archives. Once that became routine, "a structure" stopped being only an argument presented by one group. It became a reusable digital artefact. Students could download it for a class. A competing lab could test whether an interpretation held up. A software developer could treat it as training data. A drug designer could pull out a binding pocket and ask whether a candidate molecule could fit. The result was a quiet cultural shift: molecular portraits became part of a growing, searchable, inspectable commons.

The most widely recognized of these archives is the Protein Data Bank, usually called the PDB. It began in an era when "database" meant something closer to a curated library than a cloud service, and when the number of known macromolecular structures could still be counted in the dozens. The early vision was both practical and idealistic: coordinates describing atoms in a protein or nucleic acid should not be trapped in a figure; they should be deposited in a standard format so anyone can use them. As more structures accumulated, the archive became a kind of shared memory for molecular science. If you want to know how hemoglobin holds oxygen, how a kinase binds ATP, or how a receptor cradles its ligand, the PDB is where those models live.

Cryo-EM needed a parallel kind of memory, because it produces something that is related to an atomic model but not identical. A cryo-EM reconstruction is a three-dimensional density map: a volume showing where mass is likely to be, derived from many particle images. At high clarity, an atomic model can be built into that map, and the model can then be deposited in the PDB. But the map itself is not redundant. It contains the direct evidence from which the model was inferred, along with information about variable quality across the structure: crisp regions and blurry ones, stable cores and flexible appendages. Keeping only the atomic coordinates would be like keeping a transcription of a witness statement and discarding the recording. You could still quote it, but you would lose the ability to check how confident the witness sounded and what was ambiguous.

That is why cryo-EM developed its own map archive, the Electron Microscopy Data Bank, or EMDB. In cryo-EM work, EMDB is the home for three-dimensional EM reconstructions, including single-particle maps and tomograms. It complements the PDB rather than competing with it. The PDB stores atomic models: the best estimate of where atoms sit. EMDB stores the map: the reconstructed density. Together, the pair makes a claim testable. Does the model really fit the map? Are there alternative interpretations in flexible regions? Could a different modelling choice explain the same density? These are not abstract worries. They are everyday questions in a field where the human brain is excellent at seeing patterns, and where the temptation to "complete" a partially resolved region can be strong.

A third archive adds another layer, one that would have seemed extravagant in the era of expensive storage and slow internet: EMPIAR, the Electron Microscopy Public Image Archive. Cryo-EM structural biology is supported by coordinated public archives: EMDB for 3DEM reconstructions, PDB for atomic models, and EMPIAR for raw 2D image data. EMPIAR exists for the earliest, messiest, and in some ways most valuable materials: the raw micrographs and intermediate datasets from which reconstructions are built. In ordinary lab life, those micrographs are the grainy starting point, the place where beam-induced motion and contamination are battled, where good particles are separated from bad. Sharing them allows something unusually powerful: reprocessing from scratch.

Reprocessing is not a luxury. It is a way to keep science honest and to let it improve. Cryo-EM has been shaped by rapid advances in detectors, algorithms, and statistical methods. A dataset collected ten years ago might yield a substantially better map today, not because the sample changed, but because motion correction, contrast transfer function estimation, particle picking, and classification have improved. Data standards and deposition pipelines evolved to support joint map+model deposition and improved metadata and validation; EMPIAR supports sharing of raw and intermediate data to enable validation and methods development. A raw dataset in EMPIAR can become a benchmark: developers can test a new alignment method against known results, and they can report, credibly, whether it improves resolution or reduces artefacts. The archive becomes a public test set for the entire community.

The difference between "map" and "model" can sound technical, but it translates directly into trust. Imagine two labs producing two models of the same machine. If both models are deposited without their underlying maps, disagreements can become a matter of reputation or rhetoric. If both maps are deposited as well, a third party can compare the evidence. If the raw images are also accessible, the chain can be examined from the beginning: were particles mispicked, were certain classes underrepresented, did a reconstruction algorithm overfit noise, did a mask artificially sharpen features? This kind of scrutiny can be uncomfortable, but it is also how a field matures. When data are shareable, errors can be detected and corrected without relying solely on the authors' authority.

A concrete example of how archives change practice can be found in the world of method development. Much of cryo-EM's recent progress has come from software: better ways to correct drift, to estimate optical distortions, to classify heterogeneous particles into meaningful states. Those improvements need test cases. A graduate student writing a new algorithm needs datasets with known challenges: preferred orientations, conformational variability, low contrast, small particles. Publicly deposited micrographs in EM-

PIAR provide exactly that. They let developers compare old and new approaches on identical input, and they let reviewers and readers reproduce those comparisons. Without shared raw data, claims about algorithmic superiority would be mostly rhetorical.

Archives also change education. Students learning molecular biology no longer have to take a textbook diagram on faith; they can rotate a deposited model and see how subunits interlock. Students learning cryo-EM can inspect maps and discover, firsthand, that a "resolution number" is not a magic stamp. They can see that one region supports confident modelling while another is genuinely uncertain. This is an unusually healthy lesson for young scientists: knowledge has texture.

The idea of texture extends beyond cryo-EM. X-ray crystallography has long deposited models in the PDB, and the culture of validation and deposition grew partly in response to early cases where models were later found to be incorrect or insufficiently supported. The point is not that scientists are careless; it is that inference is hard and incentives can be perverse. When careers depend on striking results, the pressure to present a clean story is real. Public deposition provides a counterweight by making it easier for the community to check claims and to build on them responsibly.

Cryo-EM posed a special challenge for deposition culture because of its layered nature. What counts as "the data"? A cryo-EM experiment produces raw movies, from which corrected micrographs are computed; particles are picked; particles are refined into reconstructions; reconstructions may be sharpened and filtered; atomic models may be built and refined. Each step introduces choices. A map may be processed in ways that make it more visually interpretable but also easier to over-interpret. A model may fit a map under one refinement protocol but not under another. If only the final map is shared, the choices leading to it become hard to reconstruct. If raw images are shared as well, the choices can be revisited.

The field has responded by making deposition more systematic and by clarifying what each archive is for. EMDB policies define EMDB as a public archive for 3D EM reconstructions and tomograms and establish its role as a wwPDB core archive; the policy framework clarifies the relationship to EMPIAR for raw data deposition. This is bureaucracy in the service of clarity: the goal is to make it obvious where to find each layer of evidence, and to make deposition feel like part of normal scientific hygiene rather than an afterthought.

There is another, newer twist in the story: not all structural claims end as a single map plus a single model. As soon as researchers began combining cryo-EM with other measurements—cross-linking mass spectrometry, SAXS, NMR restraints, biochemical constraints—a new question emerged. Where do you deposit a structure that is explicitly integrative, built from multiple data types, represented as an ensemble of models, with uncertainty baked in? A conventional PDB entry is geared toward a single coordinate set with associated experimental data of a specific kind. Integrative models can include multiple representations (some atomic, some coarse), multiple compatible solutions, and a complex provenance describing how restraints were used.

Beyond single-method structures, integrative or hybrid models are increasingly archived via dedicated systems (PDB-IHM), extending the same idea of public memory to multi-technique, provenance-rich structural claims. The existence of PDB-IHM signals a maturing attitude: integrative models are not second-class, not merely illustrative cartoons in a paper. They are research products that deserve deposition, curation, and validation. That, in turn, encourages better practice. If you know that restraints and protocols will be deposited alongside the model, you have an incentive to keep them organised, to document decisions, and to think carefully about uncertainty.

Uncertainty is the thread that ties all of this together. In everyday speech, "data" sounds firm, while "interpretation" sounds optional. Molecular structure blurs that distinction. A density map is already an inference from noisy measurements. An atomic model is an inference from a map. An integrative model is an inference from multiple partial constraints. The healthiest thing a community can do is not to pretend these inferences are trivial, but to make the evidence accessible so that inference becomes a shared, checkable act rather than a private one.

This shift has had an unexpected social consequence: the community's standards become embedded in the archives themselves. Deposited entries carry metadata describing how data were collected and processed. They carry validation statistics that encourage scepticism without demanding cynicism. A reader does not have to rely solely on prose in a paper to judge whether a model is plausible; they can examine the deposited map, compare it with the model, and consult community-agreed checks. The archive becomes not only storage but a teacher of good practice.

It also becomes a catalyst for speed. When SARS-CoV-2 emerged, for example, the rapid sharing of structural data helped laboratories around the world coordinate their efforts: compare results, design constructs, interpret mutations, model interactions. This speed is not produced by any single microscope or any single lab. It depends on a background infrastructure where deposition is normal and where thousands of researchers are fluent in retrieving and using shared structures.

The deeper story is that a molecular portrait gains power when it can travel. A structure locked in a paper is persuasive to the eye, but it is inert. A structure deposited as a map, a model, and ideally raw data, becomes something closer to a scientific instrument: other people can point it at new questions. They can test new algorithms against it. They can use it to teach. They can discover where it is uncertain, and they can improve it. In a field built on inference, that kind of collective revisability is not a luxury; it is what turns impressive images into reliable knowledge.

8.4 From Portrait to Prescription

A molecular structure can be intoxicating in the way a good map is intoxicating. It promises orientation. It suggests that, somewhere in the thicket of biology, there is a surface you can touch: a pocket, a hinge, a vulnerable seam. For anyone who has watched drug discovery from the outside, this is often the imagined moment when the problem becomes solvable: find the shape, design the key, cure the disease.

Inside laboratories and companies, the mood is more complicated. A structure rarely *delivers* a medicine. It offers a chance to be less wrong. It narrows the search space. It can rescue a project that is drifting, or it can end one that is built on a false assumption. Cryo-EM has made that kind of clarity available for targets that used to be absent from the structural record: large assemblies, membrane proteins, and complexes that only make sense when multiple partners are bound at once. Those are precisely the sorts of targets that dominate modern therapeutic interest.

The first hurdle is psychological. A protein is not a single sculpture; as the ribosome makes vivid, it is often an ensemble of shapes. A drug is not just a shape-matcher; it is a molecule that must survive in blood, cross membranes, evade metabolic destruction, avoid toxicity, and achieve the right concentration in the right tissue. Between a structure and a prescription lie chemistry, physiology, manufacturing, and clinical trials. The most honest view is that structure is one kind of evidence, enormously useful but never sufficient on its own.

That said, there are moments when a new structure changes the emotional temperature of a field. One such moment came with membrane proteins. These proteins sit in the oily environment of a cell membrane, where they sense signals, transport ions, and communicate with the outside world. They are also the targets of a large fraction of drugs, precisely because they present accessible surfaces and because their activity controls so much physiology. Yet membrane proteins have long been difficult to crystallize. They often require detergents or specialized lipid environments to remain stable, and they can be reluctant to pack into the orderly lattice that X-ray crystallography asks for.

Cryo-EM helped loosen that bottleneck. A particularly vivid example is the ion channel TRPV1, a protein intimately connected with pain sensation and famously activated by capsaicin, the compound that makes chilli peppers feel hot. Cryo-EM enabled high-resolution membrane protein structures (e.g., ion channels like TRPV1) that were historically difficult to crystallize, expanding the structural basis for therapeutically relevant targets. When the TRPV1 structure appeared in 2013, it did more than provide a static picture. It offered a framework for thinking about *gating*—the channel's transition between closed and open states—and about where different modulators might bind to influence that transition. Channels, after all, are not interesting because of their mere existence; they are interesting because they flip between states, and those flips can be biased by small molecules.

This is one of cryo-EM's most practically important contributions to drug discovery: it makes conformational states more visible and therefore more discussable. A classical "lock-and-key" view assumes a fairly rigid protein that either accommodates a ligand or does not. Many modern targets operate through allostery, where binding in one place shifts the shape elsewhere, changing activity. If a protein has a rarely populated "cryptic" pocket that appears only in a certain conformation, then a single structure may miss it entirely. Multiple cryo-EM states can reveal it, not by magic, but because classification can separate the population into distinct arrangements and because the sample can sometimes be prepared to favour one state over another.

The word *druggable* captures a hard-earned realism. Not every pocket can bind a small molecule tightly enough to matter in the body, and not every binding event changes function. Structural information helps triage these possibilities. A deep pocket lined with complementary chemical groups is a good candidate. A flat, featureless surface is harder. An interface between two proteins can sometimes be targeted, but it may require a molecule larger than a typical pill-like drug, or a cleverly designed compound that wedges into a transient groove. Structures make these constraints tangible. They do not guarantee success, but they prevent certain kinds of wishful thinking.

A second example, closer to everyday pharmacology, comes from the sprawling family of G protein-coupled receptors (GPCRs). GPCRs sit in membranes and translate external signals—hormones, neurotransmitters, sensory cues—into cellular responses by coupling to internal partner proteins such as G proteins. For years, most GPCR structures were obtained either as receptors stabilized in particular conformations or as engineered constructs. Those achievements were profound, but they could leave an important question dangling: how does the receptor look *while it is actually engaging its transducer?*

Cryo-EM made receptor—transducer signaling assemblies structurally accessible, supporting mechanism-informed drug discovery beyond static receptor snapshots. A widely discussed case was the activated GLP-1 receptor (a key target in metabolic disease) in complex with a G protein, solved in 2017. What such a structure offers is not just a receptor with a ligand in its binding site, but a receptor caught in an active arrangement, gripping its intracellular partner. It turns signalling from a black box into a geometry problem: which helices swing outward, what new cavity opens on the inside, how the binding of an agonist is coupled to engagement of the G protein.

This matters because GPCR drug discovery has moved beyond the simple question of "does it turn the receptor on or off?" Modern pharmacology cares about *which pathways* are favoured: a phenomenon often discussed as biased signalling. A ligand might activate one downstream response more than another, and those differences can translate into efficacy or side effects. Structures do not directly measure signalling outcomes, but they can suggest plausible mechanisms: how one ligand stabilizes a receptor conformation that more readily couples to one partner than another, or how an allosteric modulator changes the receptor's flexibility.

Recent GPCR-focused reviews connect structural advances to drug discovery concepts such as orthosteric, allosteric, and bitopic ligands, and can frame how cryo-EM-derived interactions inform medicinal chemistry hypotheses. The terms sound forbidding, but the idea is intuitive. An *orthosteric* ligand binds to the receptor's main active site—the place nature uses. An *allosteric* ligand binds elsewhere and changes the receptor's behaviour indirectly. A *bitopic* ligand, as the name suggests, spans two regions at once, combining features of both. Cryo-EM structures of full signalling complexes can reveal not only the orthosteric pocket but also new crevices at interfaces and hinge regions that might serve as allosteric sites. They can also show why bitopic strategies might work: perhaps a ligand could anchor in the canonical pocket while reaching into a neighbouring groove that is unique to one receptor subtype, improving selectivity.

Selectivity is one of the most stubborn problems in drug design. Many proteins come in families with similar binding sites. A compound that binds one family member may bind others, producing unwanted effects. A structure can reveal subtle differences that chemistry can exploit: a slightly deeper sub-pocket, a distinctive polar residue, a movable loop that creates an opportunity for a tailored interaction. Cryo-EM expands the range of targets for which such comparisons can be made, particularly when receptors or channels are studied in complexes that stabilize relevant conformations.

Yet here, too, realism is essential. A cryo-EM map is not automatically a recipe for a drug-like compound. Drug discovery reviews emphasize that actionable cryo-EM depends on workflow realities—sample preparation, conformational heterogeneity control, ligand occupancy, throughput—not just nominal resolution. Each of these phrases points to a practical snag that can decide whether a structure is useful.

Sample preparation is often the hidden hero and the hidden bottleneck. To see a ligand bound, it must actually bind under the conditions used, and it must occupy the binding site in a substantial fraction of particles. If occupancy is low—if only some receptors are bound, or if the ligand binds weakly and comes and goes—the ligand density may vanish into the average. A medicinal chemist wants to know whether a particular chemical group forms a specific contact. A cryo-EM map may not support that level of certainty if the ligand is flexible, partially occupied, or located in a region of lower local resolution. In those cases, the structure is still valuable, but the questions it can answer must be chosen carefully.

Conformational heterogeneity, the same phenomenon that makes cryo-EM powerful for machines like the ribosome, can be a mixed blessing in drug design. If the target adopts multiple states, classification can sometimes separate them, but classification is not a free lunch. Each additional state divides the data, leaving fewer particles per reconstruction and reducing clarity. Sometimes the most useful experimental move is not computational but biochemical: engineer a stabilizing mutation, add a binding partner, use an antibody fragment, or choose a ligand that biases the receptor strongly toward a single state. These interventions can feel artificial, and they must be interpreted with care, but they can make the difference between an interpretable map and a fog.

Throughput is a less romantic constraint but no less decisive. Drug discovery often involves hundreds or thousands of compounds. A structural method that takes weeks per condition will struggle to serve as a routine screening tool. Cryo-EM has become faster, but it still involves a chain of steps that can be slowed by sample instability, preferred orientations, or the need for careful grid preparation. For this reason, cryo-EM is often used strategically: to establish a binding mode for a promising series, to confirm that a compound binds as predicted, to reveal why a modification improved potency, or to explain an unexpected failure. It can also be used earlier, as a way of deciding whether a target is tractable and where to focus chemical effort.

This strategic role can be lifesaving for projects in a very literal sense. If a compound appears to "work" in a cellular assay but binds off-target, a structure can sometimes reveal the mismatch between hypothesis and reality. If a receptor's active conformation turns out to differ from what a model predicted, a structure can redirect chemistry away from a dead end. If a mutation conferring drug resistance changes the shape of a pocket, a structure can suggest how to redesign the compound to regain binding. These are not glamorous victories. They are the daily work of reducing uncertainty.

There is also a public-health dimension to this story that is easy to overlook. Many important drugs act by interfering with molecular machines: ribosomes, channels, receptors, viral proteins. When cryo-EM brings new classes of targets into view, it expands the set of places where intervention is plausible. That does not guarantee that a safe, effective medicine will follow, but it improves the odds that researchers are aiming at a real, specific mechanism rather than a guess.

The most honest way to describe the link between a structure and a therapy is as an iterative conversation. A structure suggests a hypothesis about binding or mechanism. Chemistry tests that hypothesis by making compounds. Biology tests those compounds in cells and animals. The results feed back into new structures, perhaps capturing a new conformation, a resistance mutation, or a different ligand. Cryo-EM fits naturally into this loop because it can cope with large, flexible targets and because it can reveal multiple functional arrangements rather than insisting on a single canonical pose.

A molecular portrait, at its best, is not a trophy image pinned to a wall. It is a working document: a way to keep imagination tethered to geometry, to keep ambition tethered to evidence, and to make the invisible parts of disease and therapy concrete enough to argue about with precision.

Chapter 9

The Spike in the Spotlight

When a new coronavirus began circling the globe, one molecule quickly became the star witness: the spike protein that lets the virus enter human cells. What made the difference was not only that scientists could image it, but that they could do so fast enough, clearly enough, and openly enough to turn shape into action.

9.1 The Shape That Decided the Battle

A coronavirus is, in one sense, a minimalist. It carries no muscles to swim, no legs to climb, no jaws to bite. It is a capsule of genetic instructions wrapped in a membrane borrowed from a host cell, drifting until it collides with the right kind of cell and persuades it to open the door. Almost everything that persuasion requires is concentrated into the most conspicuous feature of the virus: the spikes that stand out from its surface like a crown.

Calling them *spikes* makes them sound like simple decorations. They are not. Each spike is a machine for three jobs that must happen in the right order and the right place: latch onto a human cell, trigger entry, and then get out of the way so the viral genome can begin its takeover. In SARS-CoV-2, the spike performs these jobs with a kind of molecular choreography: parts of it act as sensors, parts act as clamps, and part of it is a tightly coiled spring waiting for release.

For the immune system, the spike is also the obvious target. If antibodies can block the spike, the virus cannot start the infection. But there is a catch that is easy to miss if you picture a protein as a single, fixed object. A protein is more like a hand than a hammer. It flexes, pivots, opens, closes, and sometimes snaps into an entirely new posture when it does its work. The immune system does not train against an abstract label like "spike protein." It trains against a particular surface: the exact hills and grooves that happen to be exposed on the virus at the moment an antibody meets it. In that sense, immunity is partly a problem of geometry.

The spike's geometry changes dramatically during entry. Before the virus fuses with a cell, the spike is in what biologists call the *prefusion* conformation. After entry begins, the spike refolds into a *postfusion* conformation. These are not subtle variations. The postfusion form is what the machine looks like *after* it has fired. The prefusion form is what it looks like *before* firing, when the virus is still outside the cell and vulnerable to being stopped. Vaccines and many therapeutic antibodies work best when they present, or recognize, that prefusion surface.

To see why, it helps to picture what the spike must physically accomplish. The virus is wrapped in a fatty membrane, and so is the human cell. Two membranes do not readily merge; they repel and resist, like two soap bubbles pressed together. Fusion proteins are nature's workaround: they use stored energy in a coiled structure to force membranes into contact and then into continuity.

SARS-CoV-2 spike is a *trimer*: three copies of the same protein assemble into a single complex that juts out from the viral surface.

Each copy has two major functional regions. The upper part, called S1, contains the receptor-binding domain (RBD), which can grab the human receptor ACE2. The lower part, called S2, contains the fusion machinery. When the spike is prefusion, S2 is folded in a way that keeps the fusion elements tucked away and safe. When the moment comes, S2 undergoes an enormous refolding that drives the viral and cellular membranes together.

The transition is so large that it creates a trap for the immune system. Antibodies are built to recognize shape. They are astonishingly good at it: a good antibody can distinguish two proteins that differ only slightly in surface contour or chemistry. But that precision comes with a cost. If the target changes its posture, the antibody that once fit can become useless, like a key cut for a lock that has been replaced.

This is not an abstract concern. Many viruses carry fusion proteins that swing between prefusion and postfusion shapes, and vaccine history is scattered with examples where the immune system was pointed, inadvertently, at the wrong one. A particularly instructive case predates COVID-19. Respiratory syncytial virus (RSV) has a fusion protein called F that also refolds from prefusion to postfusion. For years, vaccine attempts produced plenty of antibodies, yet many did not protect well. The crucial realization was that the most potent neutralizing antibodies in natural infection recognized features unique to the prefusion form. Once the prefusion structure was determined and stabilized, vaccine design improved markedly. That earlier lesson hovered over coronavirus work like a warning: *make sure you are showing the immune system the right face.*

Coronaviruses had offered their own warning, too. After the 2002–2003 SARS outbreak and the later MERS outbreaks, researchers had learned that coronavirus spikes behave like spring-loaded devices. They can be metastable: content to sit in a prefusion pose until triggered, and then strongly biased toward collapsing into the postfusion form. The postfusion structure is not merely an alternative; it is the downhill, spent state. If you are trying to make an antigen in the lab—a purified spike protein to study or to use in a vaccine—that protein may have a tendency to slip into postfusion during production or handling. If it does, the surface presented to the immune system will be the surface of a machine that has already fired.

Why does that matter so much? Because the most useful antibodies are those that stop infection at the starting line. Neutralizing antibodies often work by blocking the RBD from binding ACE2, or by preventing the structural rearrangements required for fusion. Both kinds of interference are fundamentally prefusion interventions. Once the spike has refolded, many of the most protective sites are gone. Some are hidden; some are rearranged; some are physically far apart. An antibody raised against postfusion may still bind to the postfusion spike beautifully in a test tube and yet do little to stop a virus particle approaching a cell, because the relevant geometry on the virus is not the one the antibody was trained on.

The spike adds a second layer of geometric trickery: even within the prefusion form, it is not static. The RBD can tilt upward into an "up" state, which exposes the ACE2-binding surface, or tilt downward into a "down" state, which is more shielded. It is tempting to think of "up" as simply "ready" and "down" as "not ready," but the deeper point is that the virus can modulate which surfaces are available for antibodies to grab. A downward-tilted RBD hides key features, reducing the chance that an antibody will latch on. An upward-tilted RBD makes receptor binding possible but may increase vulnerability. The virus is balancing these constraints: it must reveal enough to infect, but not so much that it is easily neutralized.

Then there are the sugars. The spike is decorated with glycans: chains of sugars added by the host cell that produced the virus. These glycans can act like a partial cloak, blurring the protein surface and limiting where antibodies can easily land. The cloak is imperfect—antibodies *can* target protein regions between glycans, and some antibodies can even interact with glycan-associated surfaces—but it changes the landscape. Again, immunity becomes a question of three-dimensional access: where, on this moving, sugar-studded object, can an antibody physically fit and still block entry?

All of this might sound like discouragingly fine detail. In practice, it became a decisive advantage, because the spike's moving parts could be pinned down by high-resolution imaging, and those images could be translated into vaccine decisions quickly.

One of the most striking episodes early in the pandemic was how fast the spike's prefusion structure became public. The viral genome sequence was released in January 2020, and within weeks researchers were producing stabilized spike proteins and collecting cryogenic images. Daniel Wrapp and colleagues, working with Jason McLellan's group, reported a prefusion spike structure by single-particle cryo-EM in early 2020, and an early model (commonly associated with PDB entry 6VSB) was deposited on 10 February 2020. That date matters not because of numerology, but because it captures a shift in biology's tempo: a concrete three-dimensional target for vaccines and antibody discovery existed while the world was still learning the virus's name.

What did that target change? It changed what it meant to design an immune response. Without a structure, "make an antibody against spike" can mean almost anything. With a structure, the conversation becomes physically anchored. Researchers could point to the RBD and ask: which antibodies will block ACE2? They could point to hinge regions and ask: which antibodies might prevent the spike from transitioning into fusion? They could see the arrangement of glycans and ask: where are the accessible patches that an antibody can realistically reach? They could also see, crucially, that the prefusion trimer is a particular architecture—a three-part assembly whose surfaces are created not only by each subunit alone, but by the way the three subunits touch.

A useful way to think about prefusion versus postfusion is to imagine taking a photograph of a trap. The prefusion spike is the set trap: tense, ready, potentially dangerous, but also vulnerable, because its mechanism must remain exposed enough to be triggered. The postfusion spike is the sprung trap: its parts have snapped into a different arrangement, and the surfaces you see are not the surfaces that mattered when the trap was still waiting. If you train an immune response on the sprung trap, you may produce antibodies that recognize it with high affinity, and you might even feel reassured by that binding. But those antibodies may meet a real virus only to find the trap still set, presenting a different arrangement of ridges and cavities.

The same logic explains a subtle but important point about lab assays. Many ways of measuring antibodies rely on binding: does the antibody stick to the protein? Binding is necessary for neutralization, but it is not sufficient. An antibody can bind tightly to a postfusion surface and still fail to neutralize because it cannot bind the relevant prefusion surface on an intact virus in a way that blocks entry. When clinicians and immunologists talk about "neutralizing titres" rather than binding titres, they are acknowledging this geometric reality: the immune system must not merely recognize the virus; it must interrupt a particular motion at a particular time.

This is why the phrase "spike protein" can mislead the nonspecialist. It sounds like a single entity. But the immune system is encountering an ensemble of shapes sampled by the prefusion spike, and then, if entry proceeds, a largely different postfusion object. “What protein is it?” and “What shape is it in right now?” are questions that can yield different answers in terms of protection.

The geometry also helps clarify why some antibodies are so potent. Consider an antibody that targets the ACE2-binding site on the RBD. If it binds in the right orientation, it can act like a literal physical blocker, occupying the space ACE2 would need. Other antibodies bind elsewhere, away from the receptor-binding site, yet still neutralize. How? Some appear to lock the RBD in the down position, reducing receptor engagement. Others may bridge between parts of the trimer, acting like a staple that resists the rearrangements required for fusion. Still others target regions exposed only when the spike breathes into a particular prefusion microstate. These are not metaphors layered onto a cartoon; they are mechanical hypotheses grounded in shape.

There is also a reason the prefusion shape became a shared reference point across laboratories and countries. When everyone uses the same three-dimensional portrait, they can compare results directly. An antibody discovered in one lab can be mapped onto the same surface used by another lab. Mutations in emerging variants can be located precisely: not merely “in the spike,” but on a particular ridge of the RBD, or near a glycan site, or at an interface between subunits. The conversation becomes less rhetorical and more spatial. People can argue productively about whether a change will disrupt an antibody’s contact points, or whether it will shift the balance between up and down RBD states, or whether it will alter how easily the prefusion spike collapses toward fusion.

That spatial common ground also sharpened public misunderstandings. When variants began to spread, headlines sometimes implied that mutations meant the virus had acquired an entirely new spike. In reality, many variants remained recognizably the same machine, but with altered details at key surfaces. In geometric terms, the virus was sanding down some protrusions and adding others, trying to evade antibodies while preserving the ability to bind ACE2 and execute fusion. The struggle was not an all-or-nothing identity crisis; it was a continuous negotiation of surface shape.

All of this returns to the same quiet but powerful idea: protection depends on catching the virus before it crosses the threshold into the cell. The prefusion spike is the form that must do the persuading, the form that must reach out and attach. It is exposed, it is dynamic, and it is the moment of vulnerability. Once the postfusion form appears, the key step has already happened. By then, the virus is no longer asking permission at the door; it has begun to move furniture inside the house.

The remarkable thing, viewed from the scale of ordinary human life, is that this battle hinged on surfaces measured in nanometres—on whether an antibody's loops could nestle against a particular contour before that contour changed. In early 2020, as hospitals filled and countries improvised public-health measures, a parallel effort was unfolding in quiet rooms around microscopes and computers: to capture, with enough clarity to trust, the spike in the instant before it fired, and to treat that instant not as a curiosity but as the target. That choice—to aim the immune system at a specific conformation, a specific face of a moving molecule—was not merely a technical preference. It was the shape that decided the battle.

9.2 Portraits at Pandemic Speed

In the first weeks of 2020, many scientists experienced a peculiar kind of whiplash. A new virus was circulating, case numbers were rising, and the most urgent questions were brutally practical: How does it enter cells? What should a vaccine show the immune system? Which antibodies are worth chasing? Yet the answers depended on a kind of knowledge that usually arrives slowly: a reliable three-dimensional picture of a protein, precise enough to guide decisions and not just decorate a press release.

A genome sequence can be shared in minutes. A protein structure, historically, often took months or years of trial and error. The crisis did not shorten the laws of physics or the fragility of proteins. What changed was that, by 2020, a whole craft had become industrial-strength. The speed was real, but it was not reckless. It was the speed of a technique whose weak points had been identified and patched over a decade of practice: better detectors, steadier microscopes, automated data collection, and software that could turn a mountain of noisy images into a trustworthy map while keeping wishful thinking on a short leash.

One way to feel that shift is to treat public databases as time-stamped receipts. GISAID's retrospective places the early genome release on 10 January 2020 (00:41 UTC). The early prefusion spike structure associated with PDB entry 6VSB was deposited on 10 February 2020. A month, in the middle of a spreading pandemic, is not just a number. It is the difference between arguing in generalities and having a shared object to point to, rotate, measure, and test. Contemporary reporting on 19 February 2020 emphasized that the structure had been solved "in just weeks" after the genome became available. The phrasing was not marketing fluff; it captured the fact that a mature pipeline had finally met a problem that could not wait.

The drama of that pipeline begins with an uncomfortable truth: raw cryogenic images are not “portraits” in any ordinary sense. They are closer to astronomical photographs taken through a turbulent atmosphere, except the subject is smaller, the camera is bombarded by radiation damage constraints, and the specimen itself is a flexible molecule frozen mid-breath. Each exposure gives a faint, grainy projection of a particle embedded in glassy ice. The most important information is present, but it is buried in noise and distorted by the physics of imaging. A single picture of a single particle is rarely persuasive.

The crucial trick, refined long before SARS-CoV-2 arrived, is to take tens or hundreds of thousands of such particle images and combine them. The procedure is called single-particle analysis, and the name can mislead: it is not analysis of *one* particle so much as analysis of a *population* of particles, each one caught at a random orientation and, often, in a slightly different state. The end product is an average, but not a bland average. It is a three-dimensional reconstruction that can preserve crisp detail precisely because the averaging is done after the images have been grouped by viewing angle and, when possible, by conformation.

That sounds straightforward until you imagine the data in human terms. Suppose you are given hundreds of thousands of photographs of the same person taken through frosted glass. In each photo, the person is turned differently. In some, they are looking up; in others, down. Some are mid-step, some have their arm raised, some are half-occluded by a scarf. You are asked to produce a realistic three-dimensional model of their face. You cannot simply stack all the photographs and hope for the best. You need to sort them, align them, decide which differences are real and which are noise, and avoid building a face that is merely the average of incompatible expressions.

The pandemic put a spotlight on that sorting step because the spike itself refuses to sit still. As already discussed, it samples multiple prefusion states; one receptor-binding domain may be 'up' while others are 'down,' and the balance can shift. The computational work therefore had to do two things at once: increase clarity by combining huge numbers of images, and preserve truth by refusing to combine particles that did not belong together. The faster the pipeline ran, the more tempting it was to accept a single blended answer. The discipline was in resisting that temptation.

A key enabling technology was the direct electron detector. Older cameras often relied on a scintillator: electrons hit a layer that produced light, and the light was recorded. Direct detectors register electrons more directly and efficiently, raising the detective work from a dim hint to a usable signal. Just as importantly, they can record a movie rather than a single still. That matters because the specimen moves under the beam in tiny, involuntary ways. The ice can flex; the particle can drift; the stage can relax. A single long exposure smears the information. A movie lets you track and correct that drift by aligning the frames, recovering detail that would otherwise be washed out. The end result is not a cosmetic sharpening. It can be the difference between seeing a vague lobe and tracing a backbone.

Speed in 2020 also depended on something less glamorous than detectors: automation. Modern instruments can collect data for hours with minimal human intervention, moving from hole to hole on a grid, focusing, exposing, and recording. That turns the microscope into a production line for raw images. In a crisis, such throughput is not a luxury. It changes the nature of decision-making. If you can obtain enough particle images quickly, you can afford to throw away data that fail quality checks. You can be picky about ice thickness, particle distribution, contamination, preferred orientations. In older, slower workflows, a weak dataset might be all you had, and you might be tempted to interpret it generously. High throughput allows rigor precisely because it makes rejection less painful.

Yet the heart of disciplined speed lies in computation, and computation has its own dangers. When you average many noisy images, you are always at risk of seeing patterns that are not there. Human perception is famously eager to find faces in clouds. Algorithms can be equally eager to fit noise if they are not restrained. The technical term for a particularly insidious failure mode is *overfitting*: building a model that matches random fluctuations in the data rather than the underlying structure. Overfitting is not a moral failing; it is what happens when the pipeline is asked to extract maximum detail from minimal signal. The question is whether the field has good habits and checks to detect it.

By 2020, those checks were not perfect, but they were widely understood and increasingly standardized. Reconstructions were routinely validated by splitting data into independent halves, building two maps, and comparing them: if a feature appears in both halves, it is more likely to be real; if it appears only when all data are combined, it is more suspect. Researchers checked whether the 3D map supported the atomic model they built into it, rather than the model dictating what the map should contain. They inspected local resolution: were some regions crisp while others were fuzzy, suggesting flexibility? They used classification to separate distinct states rather than forcing them into a single compromise. These practices do not eliminate error, but they change its character. Instead of hidden, unquantified uncertainty, you get uncertainty with signposts.

The spike offered an immediate demonstration of why that distinction is not academic. A blurred reconstruction might still reveal that the spike is a trimer, roughly the right size, and roughly where the receptor-binding domain sits. But "roughly" is not enough to guide vaccine antigen design or antibody discovery. A neutralizing antibody can fail if it misses by a nanometre; a stabilizing mutation can backfire if it nudges a loop in the wrong direction. The most valuable maps were the ones that let researchers see not just the overall silhouette, but the placement of helices, the bending of linkers, the positions of glycans, and the subtle hinge-like motions that control access to ACE2. In other words, the crisis demanded *interpretable* detail: detail that could be tested, engineered, and compared across labs.

An anecdote from the early months of the pandemic captures the mood. Many non-specialists were surprised that a structure could appear so quickly, as if scientists had simply pointed a microscope at the virus and pressed "capture." Inside laboratories, the feeling was different. The speed came from familiarity. Teams were not inventing a method under pressure; they were applying well-worn routines with unusual urgency. Sample preparation still required care and luck. Grids still failed. Proteins still aggregated. Ice still formed too thick or too thin. But the moment a decent dataset arrived, the path from images to map was no longer a mysterious artisanal process. It was a sequence of steps that experienced groups could execute rapidly, and that other groups could scrutinize.

That scrutiny mattered because early spike structures rapidly became a shared scaffold for a huge range of work. Antibody hunters could map where candidate antibodies bound. Virologists could interpret how mutations might change receptor engagement. Biochemists could design constructs to express the spike more stably. Computational biologists could run simulations starting from a realistic shape rather than an educated guess. The structure did not answer every question, but it made many questions askable in a precise way.

At the same time, disciplined speed meant acknowledging what cryogenic reconstructions can and cannot tell you. A reconstructed map is not the virus itself. It is an inference from many images of molecules prepared in a specific way, often using a soluble version of the spike rather than the spike embedded in a viral membrane. Such constructs are necessary for many experiments, but they can differ from the spike as it exists on the intact virus. Some regions may be trimmed to help expression. Some flexible parts may be unresolved. Some conformations may be overrepresented because they survive purification better. None of this invalidates the structures; it simply defines their scope. In a crisis, scope boundaries can be forgotten by people eager for certainty. The discipline lies in stating, and remembering, that a map is a constrained view: powerful, but not omniscient.

The software side of the story also has a human dimension. High-resolution reconstructions are not produced by a single button press. They are produced by many decisions: how to pick particles, how to exclude junk, how to correct for optical distortions, how to choose masks and filters, how to classify heterogeneity, how to assess resolution. These decisions can be made responsibly, but they can also be tuned, consciously or not, toward a desired outcome. During COVID-19, the community's growing culture of sharing helped counter that risk. Depositing structures quickly into the Protein Data Bank created a common reference. In many cases, sharing of intermediate products—maps in dedicated archives, and sometimes raw images in places like EMPIAR—made it possible for others to re-run analyses and test how sensitive the result was to processing choices. Reanalysis is not an accusation; it is a form of peer review suited to a data-rich method.

It is also worth noticing what did *not* happen. There was no widespread collapse into "anything goes." In a frantic atmosphere, one might expect the field to accept lower standards and publish whatever could be made to look plausible. Instead, many groups relied on practices that had matured during the resolution revolution of the 2010s: conservative validation, careful reporting, and a willingness to show uncertainty as well as clarity. The pressure of the pandemic did not create those norms; it tested them, publicly, at scale.

When people later spoke about "pandemic speed," it was tempting to treat it as an extraordinary one-off, powered by adrenaline. Adrenaline helps, but it does not align hundreds of thousands of particle images or prevent a model from fitting noise. The deeper explanation is that a technique can be fast without being sloppy if it has become predictable: if it has standardized checkpoints, known failure modes, and tools that help practitioners distinguish signal from imagination. A crisis rewards that predictability. It lets scientists move quickly *and* remain credible, which is the only kind of speed that can support public-health decisions.

The spike structure deposited on 10 February 2020 was not the endpoint of a race; it was a starting pistol for a different kind of competition: a competition between a virus mutating through human populations and a global scientific community trying to keep its molecular picture sharp enough to guide countermeasures. The image had to be more than quick. It had to be reliable enough that, when someone made a vaccine design choice or invested in an antibody, they could say: this is not a hunch, it is a shape we can all examine.

9.3 Locking the Target in Place

A good portrait is not only about the camera. It is also about persuading the subject to hold still long enough for the picture to mean something. With the SARS-CoV-2 spike, that persuasion became a kind of molecular stagecraft. The most useful target for vaccines and neutralizing antibodies was a particular prefusion arrangement, yet the spike itself is built to escape that arrangement. Its job is to refold, force membranes together, and end up in a spent postfusion form. In other words, the protein is designed to *stop* being the thing immunologists most want to show the immune system.

That is an awkward starting point for vaccine design. The immune system can only learn from what it is given. If a vaccine delivers a spike protein that has already relaxed into an after—the-fact shape, the immune system may produce antibodies that bind nicely to that shape and still fail to block infection effectively. The solution was not merely to obtain a structure and admire it, but to engineer the protein so that, under real-world conditions—in a vial, during shipping, after a freeze—thaw cycle, at the temperature of a clinic—it would *prefer* to remain in the right pose.

Protein engineers often talk about "stability," but that word hides two distinct ideas. One is the everyday sense: does the protein stay folded, or does it fall apart into a useless tangle? The other is more strategic: among several folded shapes that are all physically possible, which one does the protein occupy most of the time? A molecule can be perfectly folded and still be "unstable" in the sense that it keeps flipping into a conformation you do not want. The spike creates both problems at once. It can be difficult to express and purify in large quantities, and it also tends to slip toward fusion-ready or postfusion-like states unless it is restrained.

The restraint came from a family of small changes to the amino-acid sequence: stabilizing mutations. The concept is deceptively simple. Proteins are chains of amino acids, and particular sequences make it easier or harder for the chain to bend in certain ways. Some amino acids, by virtue of their chemistry and geometry, act like hinges; others act more like stiff rods. A protein refolding event, such as the spike's fusion transition, requires coordinated bending and straightening across specific regions. If you can make key parts of the chain less willing to bend in the necessary way, you can raise the barrier to the unwanted transition. You do not need to weld the protein shut; you only need to bias its probabilities.

Among the most famous stabilizing substitutions are prolines. Proline is a peculiar amino acid: its side chain loops back and locks part of the backbone into a constrained angle. Insert a proline at the right location and a flexible segment becomes more rigid, less able to adopt some of the shapes required for a large refolding. Protein engineers use prolines the way a carpenter uses a brace: not to stop all movement, but to prevent the specific movement that would collapse the structure.

By the time SARS-CoV-2 appeared, this was not a brand-new idea conjured in panic. Coronavirus researchers had already been experimenting with ways to keep related spikes in their prefusion forms, and those lessons could be reused quickly. What changed in 2020 was the scale and urgency. The stabilizing logic that once served as an elegant laboratory trick suddenly became a practical necessity. If you want billions of vaccine doses, or simply want thousands of labs to run comparable antibody tests, you cannot rely on a protein that behaves differently each time it is made.

A concrete example captures how quickly "lock it in place" turned into an engineering sprint. Ching-Lin Hsieh, Jory A. Goldsmith, and colleagues (2020) described a prefusion-stabilized spike called HexaPro, named for the six proline substitutions used to reinforce it. The headline result was not only that it stayed in the desired prefusion shape, but that it expressed better and tolerated handling stress more robustly than earlier constructs: it could better survive room-temperature exposure and repeated freeze—thaw cycles. That sounds like a footnote until you imagine the logistics of a global response. Proteins are shipped across countries, stored in freezers that sometimes fail, thawed on lab benches, pipetted by exhausted technicians, and measured in assays where small inconsistencies become headlines. A sturdier protein becomes not just a nicer reagent, but a steadier foundation for the whole ecosystem of diagnostics, vaccine research, and antibody discovery.

The deeper lesson of HexaPro is that stabilization is not merely about "freezing" a protein into one pose. It is also about *manufacturability*. A spike that folds correctly and stays put can often be produced at higher yield. Higher yield means more experiments per batch, more consistent standards for serology tests, and faster iteration in design. When a protein is hard to produce, scientists spend time nursing it: changing expression systems, adjusting purification conditions, troubleshooting aggregation. When it is easier, they can spend time asking biological questions. In a pandemic, that shift is priceless.

Engineering also exposed an important, slightly uncomfortable truth: there is no single, perfectly correct spike antigen for every purpose. Different stabilizing strategies can bias the conformational ensemble in different ways. Some designs aim to keep the spike in a more "closed" prefusion state, with receptor-binding domains less frequently in the up position. Others may allow more breathing and exposure. Each choice can alter which antibody targets—epitopes—are visible. A closed trimer may present some conserved surfaces nicely while hiding others. A more open trimer may better expose the ACE2-binding region but at the cost of increased fragility or increased risk of drifting into undesired states. These are not mere technicalities; they are trade-offs between immune focus, stability, and the practical demands of production.

That is where cryo-EM becomes not just a camera but a judge. Predicting the effect of mutations on a complex protein is notoriously difficult. Intuition helps, and computer models can suggest promising sites, but proteins surprise their designers. A mutation intended to stiffen a hinge can create a new wrinkle elsewhere. Two mutations that each seem benign can interact to destabilize a distant region. A construct can look excellent in a biochemical assay and still adopt an unintended architecture that only becomes obvious when you see it in three dimensions.

Cryo-EM provides a decisive check because it answers the question designers actually care about: *what shape did you end up with?* It can reveal whether the protein is mostly prefusion, whether it is a mixture of states, whether parts are disordered, whether key regions are missing from the map because they are too flexible. It can show whether the trimer is intact or whether it tends to fall apart. It can also reveal heterogeneity: perhaps most particles are in the desired state, but a significant minority have slipped into another. For an engineering project, that minority is not noise; it is a warning sign that the construct may behave unpredictably in manufacturing or storage.

There is a social dimension to this judging role. During COVID-19, many groups around the world were designing spike constructs, producing antigens, and proposing modifications. A construct that could be validated by a widely trusted imaging method had an advantage: it was easier to adopt by others. Cryo-EM maps and models gave a shared language for discussing design choices. Rather than debating a sequence on paper, researchers could debate the presence or absence of a loop, the angle of an RBD, the clarity of a helix. The argument could be anchored in something visible.

There was also an institutional dimension that is easy to overlook amid the scientific heroism. Some early work on prefusion spike constructs included intellectual property disclosures. That is not scandalous; it is a normal feature of the interface between academic research and real-world products. But in the context of a pandemic, it highlights that "locking the target" is not only a molecular choice. It is also a decision about what becomes a shared tool and what becomes a proprietary component. The same stabilization trick can be a public reagent in one context and a licensed technology in another. These arrangements can influence which constructs spread fastest, which become standard in diagnostic kits, and which are used as reference materials across laboratories. The spike, in other words, is not only a biological object. It is a technological platform embedded in human systems of credit, funding, and access.

Engineering the spike also sharpened thinking about what it means to "imitate the virus." No vaccine antigen can replicate the full context of an intact virus particle. A soluble spike lacks the viral membrane. It may lack certain segments. Its glycans may differ depending on the cell line used for production. Yet an engineered antigen can still succeed brilliantly if it presents the immune system with the right geometric cues. The point is not perfect mimicry. The point is strategic mimicry: presenting the surfaces that elicit protective antibodies, while avoiding shapes that distract the immune response toward irrelevant targets.

A useful way to picture this is to imagine teaching someone to recognize a person from photographs. If you always show them the person wearing a helmet, they may learn the helmet, not the face. If you show them the face at the wrong angle, they may fail to recognize the person in real life. The goal is to choose a representation that captures identity in a way that transfers. Stabilizing the prefusion spike is a way of choosing the "angle" that transfers best to the moment when the virus is trying to enter a cell.

The final twist is that "locked" does not mean "dead." A well-designed stabilized spike is often still dynamic in small ways. That is part of its realism. Antibodies elicited by vaccination must cope with a target that breathes. Too rigid a design could, in principle, present an unnatural surface, eliciting antibodies that fit only the engineered version and not the viral one. The art is to block the catastrophic refolding into the postfusion state while retaining enough native-like features that the immune system learns something transferable.

HexaPro's appeal, then, was not that it created a perfect statue. It was that it shifted the balance: more protein in the right shape, for longer, at higher yield, with less fuss. In a laboratory this saves time and reduces ambiguity. In a pandemic it changes the economics of speed. A fragile antigen turns every experiment into a bespoke event. A robust antigen turns experiments into routines, and routines can be scaled.

By the end of 2020, the phrase "the spike" had entered everyday language, as if it were a single, simple object. In laboratories, it had become something slightly different: a family of engineered proteins, tuned for particular uses, all judged by whether they kept faith with a fleeting prefusion geometry. The lock was not a padlock but a bias built into the chain itself, subtle enough to preserve function for the immune system, firm enough to resist the protein's own impulse to snap into its final form.

9.4 From Archive to Action

A pandemic has its own tempo. It speeds up ordinary life, compressing months of decision-making into days. Science, however, cannot simply will itself to accelerate. Reagents still need to be made. Experiments still fail. Data still need to be checked. One of the quiet achievements of the COVID-19 response was that an older, slower rhythm of molecular research adapted without breaking: the key products of the work were not kept as private trophies on laboratory servers, but released as standardized digital objects that other people could immediately use.

The practical effect was easy to see from the outside. A research group in one country could design an antibody-binding experiment using a spike structure produced in another. A vaccine team could encode a stabilized spike sequence while consulting public coordinates for its prefusion shape. A computational group could model how a new variant mutation might change an antibody epitope without waiting for a phone call or a PDF. The deeper effect was cultural. The day-to-day machinery of archiving—depositing structures, maps, and sometimes raw images in public repositories—began to resemble public infrastructure. Like roads or power lines, it worked best when it was maintained before anyone desperately needed it.

There is a reason those deposits mattered beyond abstract openness. Cryo-EM does not produce a single, inevitable result. It produces a chain of inferences. The raw data are noisy movies of particles in vitreous ice. Those movies are processed into corrected images, then sorted, aligned, and combined into a three-dimensional map. An atomic model is then fitted into that map, refined, checked, and finally presented to the world as a set of coordinates. Each step has choices, and choices can influence outcomes. The reliability of a structure is therefore not only a property of the microscope; it is also a property of the surrounding culture: how quickly results are shared, how transparently they are validated, and how easily other experts can inspect the same evidence.

Public repositories made that inspection routine. The Protein Data Bank (PDB) has long served as a central library of atomic models. For cryo-EM, the Electron Microscopy Data Bank (EMDB) stores the reconstructed maps that models are built into. And the Electron Microscopy Public Image Archive (EMPIAR) goes one step closer to the source by storing raw image data. Each repository serves a distinct purpose. A PDB file lets you measure distances, map mutations, and design molecules against a target. An EMDB map lets you judge whether the model was fitted sensibly, or whether it is leaning on imagination in parts where the data are weak. EMPIAR enables something rarer and more demanding: independent reprocessing, where another group can take the same raw images and see whether a similar reconstruction emerges.

That last capability became surprisingly important, because transparency is powerful and complex at the same time. EMPIAR documentation notes explicitly that different processing choices can yield different reconstructions from the same raw data. This is not a scandal; it is a feature of working near the limits of signal and noise. If a molecule is flexible, some subsets of images will represent one state and other subsets another. If the images are weak, algorithms can be nudged by masks, filtering choices, or classification strategies. Open raw data therefore do not create a single "final answer." They create a space for informed disagreement, improvement, and convergence. In the middle of a crisis, that is exactly what scientific credibility needs: not the pretense that ambiguity has vanished, but a shared ability to examine where ambiguity lives.

The archives also supplied something more mundane but equally valuable: timestamps. In ordinary times, scientific progress is narrated by publication dates, which can lag behind discovery for months. Depositions in public databases provide a more granular record of when an object became available to the community. During COVID-19, those dates became part of the story. PDB and related metadata recorded when key spike structures appeared, turning what might otherwise feel like a vague recollection of "early 2020" into a traceable timeline. That traceability matters when people later ask hard questions about preparedness: Which tools existed when decisions were made? When did a particular structure become accessible to the wider world? When did variant-specific information enter the public domain?

A simple chain of custody illustrates the new tempo. Genome sharing through GISAID defined the problem in letters: a sequence that could be copied, compared, and translated into protein. Structural sharing through PDB and EMDB turned that sequence into geometry: a shape that could be targeted, stabilized, and used to interpret antibody binding. Raw-data sharing through EMPIAR added the final ingredient: the capacity for others to test the interpretation, try alternative processing, and extract additional information that the original team may not have pursued under time pressure. Sequence data enabled antigen definition; structural data enabled mechanistic interpretation and target engineering. Together, they formed a two-archive pipeline that linked early surveillance to actionable design.

A concrete example of how far that pipeline could extend is provided by intact-virion datasets. Early in the pandemic, many spike portraits were obtained from soluble spike constructs: proteins engineered to be easier to produce and image. Those constructs were indispensable, but they raised a natural question. How similar is the recombinant spike to the spike as it stands on the virus surface, embedded in a membrane and surrounded by other viral proteins?

One way to approach that question is to image the virus itself. Intact virions are more challenging specimens: they are heterogeneous, they can be damaged, their spikes can be variably distributed, and the background is messier. Yet they offer a closer view of the target in its native setting.

When raw data from intact-virion studies are deposited, they become seeds for further work. EMPIAR-10492 is one such entry, providing raw cryo-EM data underpinning reconstructions of spike proteins on intact SARS-CoV-2 virions. The immediate benefit is reproducibility: others can verify that the reported distributions and conformations are supported by the images. The longer-term benefit is methodological. As algorithms improve, old raw datasets can yield new insights. A classification method that was not robust in 2020 may become reliable in 2022, allowing a reanalysis that teases apart subtle states. In that sense, EMPIAR turns a one-off experiment into a reusable resource, keeping the scientific record alive rather than sealed.

This ability to revisit data is not a luxury. In an emergency, early results are often produced under constraints: limited time, limited reagents, incomplete knowledge of the specimen's quirks. Later, as understanding improves, the same data can be reinterpreted more carefully. Open archiving makes that reinterpretation communal rather than proprietary. It also prevents a familiar failure mode in fast-moving science: a result becomes influential, cited, and assumed, yet no one outside the original group has the means to re-examine the underlying evidence. Raw-data deposition does not guarantee re-examination, but it removes a major barrier.

The archives also influenced the economics of collaboration. In pre-digital scientific culture, collaboration often required personal trust, established networks, and the slow exchange of physical materials. During COVID-19, those modes still mattered, but public digital artefacts reduced dependence on them. A graduate student in a small lab could download a structure and start asking sophisticated questions without needing an introduction to an elite group. A pharmaceutical team could cross-check a reported epitope against a public model. A computational group could build a mutation map overnight. The result was not an egalitarian utopia—expertise, equipment, and funding still shaped who could do what—but the floor rose. Access to key molecular portraits became less a privilege and more a baseline expectation.

There was, inevitably, tension between speed and polish. Depositing a structure quickly can mean releasing a model with unresolved regions, uncertain glycan placements, or flexible loops that are difficult to interpret. Some observers treat these imperfections as flaws. In practice they can be honest signals. A missing loop in a model often means the loop was too mobile to resolve clearly, not that the researchers were careless. Public access lets others judge that for themselves. It also encourages incremental improvement: an early structure can be good enough to guide initial vaccine antigen choices and antibody discovery, and later structures can refine details as methods and samples improve.

A historical anecdote helps put this in perspective. The PDB began in the 1970s as a modest archive when protein structures were rare and precious, each one the product of years of work. Depositing a structure then had a ceremonial quality. Over time, the archive became less a museum and more a utility. By the 2010s, before SARS-CoV-2, large numbers of structures were deposited routinely, and a generation of biologists grew up treating "download the structure" as an ordinary first step. COVID-19 did not invent this openness; it stress-tested it. The system held, and because it held, the scientific response could behave less like isolated heroics and more like coordinated engineering.

The link between openness and translation was especially clear in vaccine work. Vaccine-development literature tied stabilized prefusion spike design to rapid translational action through the idea of prototype pathogen preparedness: having a playbook for a viral family so that, when a new member appears, the steps from sequence to antigen to clinical testing can proceed quickly. Kizzmekia S. Corbett, Darin K. Edwards, and colleagues (2020) emphasized how prior coronavirus research and prefusion-stabilization strategies enabled rapid mRNA vaccine antigen selection. The point is not that an archive can replace experimentation. It is that shared molecular knowledge reduces the time spent rediscovering what is already known and focuses effort on what is genuinely new.

The result was a feedback loop that would have been difficult to sustain in earlier eras. A sequence appears in a genome database. A stabilized antigen is designed. A structure is deposited. Antibodies are isolated and mapped onto that structure. Variant mutations are interpreted in structural terms and compared against antibody panels. Improved antigens are engineered and checked. Each step produces digital artefacts that can be shared widely and immediately, allowing the loop to tighten. In public-health terms, that tightening translates into earlier warnings about immune escape, quicker adaptation of assays, and a clearer rationale for vaccine updates.

The most hopeful aspect of this culture is also the most demanding. Openness asks the community not only to share results, but to share uncertainty. A structure deposited in PDB is not a declaration of perfection; it is an invitation to use, test, refine, and sometimes correct. Raw data deposited in EMPIAR are not a guarantee that everyone will agree; they are an acknowledgement that judgement is part of the method and that judgement should be auditable. In a moment when misinformation flourished alongside genuine discovery, that auditable chain—from images to maps to models, with public checkpoints in between—offered a rare kind of reassurance. Not that science always moves smoothly, but that its claims can be traced back to evidence that others are allowed to see.

When the spike became a global obsession, it did not remain confined to laboratory notebooks. It became part of a shared digital commons: coordinates that could be copied endlessly without being used up, maps that could be reinterpreted as tools improved, and raw images that could be reprocessed by people who had never met the original data collectors. For a virus that spreads by replication, it is fitting that one of the most effective responses relied on a different kind of replication: the rapid, faithful copying of knowledge.

Chapter 10

Beyond the Portrait: Life in Its Native Crowds

Cryo-EM did not end with beautiful isolated structures; its next frontier begins where order breaks down. Here the reader meets restless molecules, elusive tiny targets, and the ambition to image molecular neighborhoods inside cells, before arriving at a future in which algorithms and automated microscopes may learn to steer discovery together.

10.1 The Molecules That Wouldn't Hold Still

The temptation, when a cryo-EM reconstruction looks soft or strangely incomplete, is to blame the hardware: a microscope slightly out of tune, a detector not quite doing its job, a laboratory draft that warmed the specimen. Often the trouble is simpler and more exasperating. Biology is lively. The sample may be perfectly vitrified and the microscope perfectly aligned, and the molecules will still refuse to behave like good subjects for portraiture.

Two kinds of restlessness cause most of the mischief. One is physical: how particles choose to lie in the thin film of ice. The other is internal: how a protein flexes, breathes, or rearranges itself, so that the microscope is asked to average together not a crowd of identical copies but a crowd of relatives, each caught mid-gesture.

Cryo-EM single-particle analysis works by collecting many 2D projections and combining them into a 3D map. It is easiest to picture as trying to infer the shape of an unfamiliar object by photographing it from every direction. If the object is rigid and you have enough evenly distributed views, the missing information shrinks away as more images arrive. But if most photographs come from one direction, the reconstruction becomes lopsided: crisp where the data are plentiful, vague where the object has rarely been seen. And if the object itself changes shape while you photograph it, even a perfect spread of angles can blur into a misleading average.

The first problem has a name that sounds like a personality trait: *preferred orientation.* Many particles, once adsorbed at the air—water interface or on a support film, settle into just a few poses. A flat, disk-like complex may land like a coin on a tabletop. An elongated complex may lie like a log in a stream, nearly always presenting its broadest face to the beam. Even proteins that are not obviously flat can have sticky patches that favor one side against the surface, and the ice in a grid hole is not a featureless lake. It has boundaries, gradients in thickness, and interfaces that can behave like selective traps.

Preferred orientation is not merely an aesthetic annoyance; it creates a specific kind of ignorance. In a 3D reconstruction, each projection image contributes information to a region of Fourier space, a mathematical bookkeeping system that tracks which spatial frequencies have been measured. A well-distributed set of orientations fills that space evenly, giving the algorithm the ingredients it needs to build a balanced volume. A biased set leaves wedges and gaps, producing direction-dependent resolution: one axis sharp, another smeared. The map may look acceptable at first glance and still be missing entire classes of features, like a face reconstructed from only profile photographs. The danger is subtle because the reconstruction software will do its best with what it has. It will not announce, in plain language, that the particle was never really seen from above.

The second problem, *conformational heterogeneity*, is more honest: it reflects the reality that many molecular machines do not have a single pose worth calling *the* structure. Ion channels open and close. Ribosomes swivel. Polymerases clamp onto nucleic acids and release them. Even proteins that seem like sturdy scaffolds often contain hinges and loops that shift as they bind partners or respond to chemical cues. A cryo-EM grid can freeze these states indiscriminately. If the dataset contains a mixture and all images are forced into one reconstruction, the result is a compromise that may correspond to no molecule that ever existed. It can also disguise the very movements that make the system interesting.

The story of the ribosome provides a concrete sense of how these issues play out in practice. Ribosomes are abundant, large enough to image well, and central to life, which made them early favorites for cryo-EM. Yet they are also dynamic machines with parts that rotate and ratchet as they translate genetic code into protein. As ribosome datasets grew in size and quality, researchers found that the familiar act of averaging could erase functionally important differences. A dataset might contain ribosomes caught before translocation, during translocation, and after, as well as ribosomes carrying different accessory factors. The algorithms could be asked to produce one map, but the ribosome refused to be reduced to a single still. Better reconstructions required a willingness to admit multiplicity and to let the data separate into states.

Preferred orientation can also masquerade as biology. Imagine a protein complex that really does change shape, but only a subset of shapes sticks to the grid in a favorable way. The dataset then becomes doubly biased: it over-represents particular views and particular conformations. A naïve analysis might conclude that the favored conformation is the predominant one in solution. The grid has quietly edited the movie.

This is where the craft of sample preparation and the discipline of computational triage meet. An experienced cryo-EM practitioner learns to treat a first dataset less as a final verdict than as a clue to what the grid is doing to the sample. Tiny adjustments in chemistry and handling can reshape the entire distribution of orientations and conformations.

One of the simplest levers is the support film. Traditional holey carbon grids leave open windows of vitreous ice suspended over holes. Particles in those holes often encounter the air—water interface, a place that can denature proteins, concentrate them, or pin them in favored poses. Adding a continuous support film—amorphous carbon, gold, or graphene-based supports—can change the adsorption landscape. A film may provide a gentler landing surface, keep particles away from the interface, or alter which side of the molecule prefers contact. The choice is rarely neutral. A support can introduce its own bias by favoring different orientations, and it can also increase background signal. Yet for difficult samples, a support film is sometimes the difference between a dataset that is reconstructable and one that never escapes preferred orientation.

Ice thickness is another quiet dictator. If the ice is too thin, tall particles are squeezed and may be forced into particular poses, like a book pressed between two panes of glass. If the ice is thicker, particles have more freedom to tumble, but thick ice reduces contrast and can overlap particles along the beam direction, complicating analysis. Even within a single grid square, thickness can vary from the center of a hole to the edge, and particles may behave differently in each region. Skilled data collection often involves hunting for the Goldilocks zone: thin enough for good signal, thick enough to permit variety.

Buffer conditions add still more degrees of freedom. Salt concentration, pH, glycerol, mild detergents, and specific ligands can all alter how a molecule behaves at interfaces and how stable it is during blotting and freezing. Sometimes a tiny amount of detergent reduces sticking and improves angular coverage. Sometimes binding a substrate or inhibitor reduces internal flexibility, turning a blurred ensemble into a well-defined state. At other times, the goal is the opposite: to preserve a range of conformations because the biological question lives in the movement. In those cases, the computational work must shoulder more of the burden.

One practical trick for preferred orientation is so direct that it feels like cheating: *tilting* the specimen during data collection. Instead of keeping the grid perpendicular to the beam, the microscope stage is tilted by a set angle, forcing the camera to see the same particles from a different direction. If particles insist on lying flat, tilting converts that flatness into a usable variety of projections. The price is paid in geometry and noise. Tilted images pass through more ice, lose contrast, and can suffer from more beam-induced motion and more complicated defocus gradients across the field of view. Tilt is therefore a remedy with side effects, used when the alternative is missing views so severe that a 3D map cannot be trusted.

All of these experimental adjustments share a philosophy: make the data less biased before asking computation to do miracles. Yet computation has become remarkably good at sorting out what biology and preparation have mixed together, provided the dataset contains enough information to begin with.

The first computational checkpoint is usually *2D classification*. After particles are picked from micrographs and roughly aligned, the algorithm groups similar images into class averages. These averages are not merely prettier pictures; they are a reality test. If the dataset contains only one view of the particle, the 2D classes will all look like slight variations on that view. If there is a good spread of orientations, classes will show distinct projections: top-like, side-like, oblique. 2D classification also reveals heterogeneity bluntly. A flexible domain may appear sharp in some classes and smeared or absent in others. A bound factor may show up as an extra blob in a subset. Contamination and broken particles often segregate into their own ugly categories. The researcher's job is to decide what to keep without discarding rare but real states. That decision is part technical, part biological judgment, and it is easy to see why the field values careful skepticism.

Once a plausible 3D model exists, *3D classification* becomes the main tool for separating conformations. Here the algorithm assigns each particle image to one of several 3D classes, each class representing a distinct map. The classes might correspond to different ligand-bound states, different assembly compositions, or different conformational snapshots. When the separation succeeds, it is deeply satisfying: features that were smeared in the averaged map snap into focus in one class or another, like a crowd parting into smaller groups where each person's outline is clearer.

But classification can also mislead. If preferred orientation is severe, the algorithm may invent classes that compensate for missing views rather than real structure, or it may merge distinct states because the available projections cannot distinguish them. If the molecule changes continuously rather than switching between discrete shapes, forcing it into a small number of classes can create artificial boundaries, as if a smooth gesture were chopped into a handful of poses. The line between genuine heterogeneity and algorithmic convenience is not always obvious.

This tension has pushed method development toward describing variability more faithfully. A major change in mindset has been to treat heterogeneity not only as a nuisance to be eliminated but as a quantity to be measured. Approaches such as multi-body refinement in RELION (Nakane and Kimanius, 2018) model a complex as several rigid bodies that move relative to each other, capturing hinge-like motions without insisting that the entire particle is one solid block. Other approaches, such as 3D variability analysis described by Punjani and Fleet (2021), aim to identify dominant modes of change in the dataset, revealing which parts of the molecule vary together and which stay put. Neural-network-based methods, including cryoDRGN (2019), go further by learning a latent space of structures—a compressed description of how the density can morph across the dataset. These methods have opened a door to richer stories, but they also ask for humility. When an algorithm is powerful enough to describe subtle differences, it is also powerful enough to mistake noise for meaningful motion if validation is lax. The old warning about overfitting does not disappear; it becomes more psychologically tempting, because the output can look like a plausible movie of molecular life.

A concrete example of heterogeneity that captured public attention arrived during the COVID-19 pandemic. The coronavirus spike protein is not a static club; it is a spring-loaded machine that exposes and hides receptor-binding surfaces as it searches for host cells. Cryo-EM revealed spikes with receptor-binding domains in different positions, including the now-familiar "up" and "down" arrangements that influence how the virus engages ACE2 and how antibodies recognize it. Those reconstructions were not merely technical triumphs; they helped guide the design of stabilized spike variants used in vaccines, where particular conformations were favored to present the immune system with an effective target. The lesson was not that cryo-EM had finally conquered motion, but that motion could be mapped and, in a sense, engineered around.

Still, for every celebrated case, there are many more where the molecule's restlessness collides with the grid's biases in an unglamorous way. A protein might have a flexible tail that flaps in solution, and on the grid that tail may stick to the surface, creating preferred orientation and also artificially immobilizing one part of the molecule. Another complex might exist in a mixture of assembled and partially disassembled forms, and the blotting process may shift the equilibrium. What looks like conformational heterogeneity could be damage; what looks like damage could be a real intermediate. The microscope does not label its own artifacts.

A useful way to think about the problem is to separate three questions that can otherwise become tangled. First: does the dataset contain enough distinct views to support a trustworthy 3D reconstruction, or is preferred orientation undermining it? Second: assuming the angular coverage is adequate, is the molecule internally consistent, or is it a mixture of states? Third: if it is a mixture, are the states discrete, continuous, or both? Each question has different remedies, and confusing them can waste months. Tilt can help the first but does nothing for the second. Binding a ligand can help the second but can also erase the biological dynamics one hoped to see. A more flexible computational model can help the third but may also produce seductive, hard-to-validate narratives.

None of this would matter so much if cryo-EM were a purely descriptive art. It is not. These reconstructions feed into claims about how enzymes work, how receptors signal, how antibiotics bind, how mutations cause disease. If a map averages away a critical loop position, a drug-design effort may chase the wrong pocket. If a reconstruction reflects an orientation bias that makes one domain look weaker, a researcher might wrongly conclude that the domain is disordered in solution. Getting the portrait right is not vanity; it is the difference between a mechanism that holds together and one that collapses under the weight of later experiments.

The stubbornness of molecules has, in a way, improved the field's honesty. Early in the single-particle era, it was easier to imagine that every macromolecule had one structure waiting to be revealed if only the microscope were good enough. The routine encounter with heterogeneity and preferred orientation has forced a more realistic view: a cryo-EM map is an inference from a dataset collected under specific physical conditions, and it comes with a personality shaped by interfaces, ice, and internal dynamics. When the molecules will not hold still, the solution is rarely a single clever trick. It is an accumulation of small choices—a different grid, a slightly different buffer, a deliberate tilt, stricter classification, a more expressive variability model—each one aimed at coaxing the dataset closer to what the biology is truly doing.

There is also a quieter emotional shift that comes with experience. At first, heterogeneity feels like failure: the data refuse to converge, the map refuses to sharpen. Over time, it can begin to feel like contact with something real. Living systems do not run on a single frozen pose. They run on probabilities, on motions that are constrained but not dictated, on parts that sample options until chemistry and chance select a path. Cryo-EM, when it succeeds against restlessness, does not just deliver a better still image. It delivers evidence that the still image was never the whole point.

10.2 Where Small Proteins Vanish

A strange thing happens as the targets of cryo-EM get smaller. The method that can make a ribosome look almost architectural, with grooves and ridges you could imagine running a fingertip along, begins to hesitate. The images turn coy. In the raw micrographs, the particle is no longer an obvious object with a recognizable outline; it becomes a suggestion, a small local disturbance in a field of mottled grey. The molecule has not changed, of course. The microscope has. Or rather, the relationship between signal and background has shifted until the protein is no longer an easy visual fact.

Cryo-EM sees by contrast: by tiny intensity variations in an image that arise because electrons have been scattered by the specimen and because the electron wave has been shifted in phase as it passes through. When a particle is big, it nudges many electrons and leaves a noticeable fingerprint. When it is small, it nudges fewer, and its fingerprint begins to resemble the random texture produced by everything else the electrons pass through: the thin film of frozen water, the support, the inevitable contamination, and the camera's own noise. At some point the particle does not merely look faint; it becomes hard to find and even harder to orient.

There is a particular cruelty in this. In principle, a small protein should be simpler than a large molecular machine. Fewer moving parts, fewer possibilities. In practice, the small protein often places heavier demands on every step of the cryo-EM inference chain. Particle picking becomes uncertain, because the thing you want to pick barely stands out. Alignment becomes unstable, because the algorithm tries to match one noisy speck to another and can be seduced by coincidences in the noise. Classification becomes ambiguous, because the data do not clearly support one pose over another. Even the act of deciding whether you have a dataset worth collecting becomes harder, because the early micrographs do not give the reassuring satisfaction of seeing your molecule plainly.

The core reason is simple enough to explain without equations. A protein is mostly light atoms. Electrons passing through it are deflected only slightly, and the number of deflections grows with the amount of material in the beam's path. That is why size, more precisely molecular weight, becomes such a blunt boundary for single-particle cryo-EM. A larger particle produces a stronger signal. A smaller particle produces a weaker one. The background, meanwhile, does not shrink in sympathy. The protein's signature slides down toward the level of everything you would prefer to ignore.

The practical consequence is that the problem is not only *imaging* but *recognizing*. If you cannot reliably say which pixels belong to your particle and how it is oriented, you cannot average in a meaningful way. Averaging is powerful, but it needs an anchor: a way to align the images so that like is added to like. With large particles, the anchor is obvious: strong features guide the alignment. With very small proteins, the anchor can be so weak that alignment starts to drift. The algorithm can align noise to noise and still report success, because it has done something mathematically consistent even if biologically meaningless. This is one reason small-protein work has a reputation for being both technically demanding and psychologically punishing: the output can look plausible long before it is trustworthy.

There is also a second, subtler issue. Many small proteins are not conveniently lumpy. They can be roughly spherical, compact, and internally repetitive at the resolutions where alignment begins. A particle that looks like a smooth pebble from most angles is hard to orient even in a good photograph. Cryo-EM often begins with images that are not much better than silhouettes, so a small globular protein can be hard to align for the same reason a featureless ball is hard to orient in space. The information is there in principle, especially at high resolution, but you need correct alignment to reach that regime. It is a circular problem: to align well you need signal, and to gain signal you need to align well.

These difficulties help explain why the field's celebrated targets have often been big. Viruses, ribosomes, chaperonins, large membrane complexes: they are generous scatterers. When cryo-EM began its modern surge, it was natural to showcase the structures that best rewarded the method. Yet many of the molecules biologists and clinicians care about are much smaller: enzymes, regulatory domains, soluble receptors, signaling proteins that act as molecular switches. Some are well under 100 kilodaltons, and many are closer to the mass of a small household object than to a cathedral. They still matter enormously to physiology and medicine, but they do not announce themselves to the microscope.

One way to appreciate the challenge is to imagine a busy auditorium where you are trying to recognize a quiet voice. You can improve the microphone. You can dampen echoes. You can ask the crowd to settle down. But if the speaker whispers, at some point the task becomes as much about clever listening strategies as about raw acoustic power. Cryo-EM has developed its own versions of each of these tactics.

A striking hardware-oriented example arrived with the Volta phase plate. The phrase sounds like a component in a science-fiction engine, but the idea is beautifully concrete. In many cryo-EM images, especially at the low electron doses needed to limit radiation damage, most of the useful information is carried not in changes of amplitude but in changes of phase of the electron wave. The trouble is that detectors record intensity, not phase. Traditional imaging tricks convert some phase information into detectable contrast, but they do so inefficiently for weak phase objects such as small proteins in thin ice. A phase plate is designed to help: it modifies the electron wave in the back focal plane of the microscope so that phase differences are turned into stronger intensity differences in the image. The result, in friendly terms, is that the molecule's whisper becomes easier to hear against the room's murmur.

In 2017, a notable demonstration used the Volta phase plate to determine the cryo-EM structure of hemoglobin at 3.2 Å. Hemoglobin, at about 64 kDa, had long been a familiar molecule from other methods and from textbooks, and that familiarity is part of what made the result resonate. The achievement was not simply a new picture of an old protein; it was a signal that contrast enhancement could push cryo-EM into territory where it had often struggled. The work also underscored a theme that recurs across the field: success rarely comes from a single gadget. It tends to be ecosystem-wide, involving optics, detectors, energy filters, specimen preparation, and computation moving forward together until a new regime becomes practical.

Even so, it would be misleading to suggest that a phase plate is a universal key. Phase plates bring their own complications, including stability and tuning, and they do not magically remove the alignment problem. They can make particles more visible, but visibility alone does not guarantee that a dataset will align cleanly and yield a reliable high-resolution map. For many small targets, the most effective strategy has turned out to be less about changing the microscope and more about changing the *specimen*.

A blunt but often successful idea is: make the target effectively larger. If molecular weight is a form of visibility, then adding mass can provide a handle for alignment. The key is to add mass in a way that is rigid and well-behaved, so that the added components do not simply introduce new flexibility and new ambiguity.

Nature already provides a convenient class of mass-adding tools: binding partners. Antibodies and their smaller cousins, nanobodies, can grip a protein tightly and add a substantial, structured bulk. The resulting complex is easier to pick and align, and it can also stabilize particular conformations. In membrane-protein work, binding partners have been used not only to add mass but to lock fragile receptors into a state that would otherwise be fleeting. The gain is clarity; the risk is distortion. A binder can hold the molecule in a pose that is rare or even non-native, or it can prevent motions that are part of the protein's real life. The extra mass is not neutral decoration. It is an intervention.

Engineered scaffolds offer another route. Researchers can design protein frameworks that present the small target at a fixed position, like a jewel set into a rigid mount. In principle, the scaffold supplies strong features for alignment while the target remains visible in a consistent location. In practice, the engineering must be delicate. The connection between scaffold and target must be stiff enough to prevent wobble but gentle enough not to deform the target. A scaffold that behaves beautifully in a design program can still flex in solution, and cryo-EM is merciless toward flexibility: if the target moves relative to the scaffold, the averaged map can show a sharp scaffold and a smeared or vanishing target, the very problem the scaffold was meant to solve.

Oligomerization is a third tactic: persuading a protein to assemble into a larger complex by design. If several copies can be linked into a symmetric arrangement, the effective mass grows and the number of views can increase because symmetry provides multiple equivalent orientations. Yet oligomerization also carries interpretive hazards. The engineered interfaces may impose constraints that do not exist in the native monomer, and the new assembly may favor conformations that are artifacts of the design. When the goal is simply to obtain *a* structure, these compromises may be acceptable. When the goal is to understand how a protein behaves in its usual cellular setting, every added constraint must be treated with suspicion.

Because these interventions can alter the biology, small-protein cryo-EM often ends up being a conversation between methods rather than a single-method triumph. Hybrid workflows have become common, where cryo-EM is combined with prior structural knowledge from other approaches. A crystal structure or NMR model might provide a plausible starting point for interpretation, or a known fold might guide how a low-resolution density is segmented. The benefit is that one does not ask cryo-EM to conjure every detail from a near-invisible signal. The danger is also clear: prior knowledge can become a crutch, or worse, a self-fulfilling prophecy. If you expect the protein to look like a known structure, you may be too willing to accept a reconstruction that vaguely resembles it. The discipline is to use external information as a guide while still demanding that the cryo-EM data justify each claim.

On the computational side, the small-protein problem shows up as a battle against low signal-to-noise ratio. In a high-SNR dataset, alignment is a robust procedure: many different random starting points converge to the same answer because the data strongly constrain the orientation of each particle. In a low-SNR dataset, alignment becomes a landscape with many shallow valleys. Different initial conditions can lead to different solutions, and the algorithm can settle into a local optimum that is consistent with noise patterns rather than true particle features. This is one reason small-protein projects are often hungry for enormous datasets: more particles can help, but only if enough of them are picked correctly and aligned correctly in the first place.

The picking step itself becomes fraught. For a large complex, it is possible to draw circles around particles by eye, train a picker on clear examples, and trust that most picks will be real. For a small protein, a micrograph can contain thousands of features that look particle-like: ice contamination, carbon edges, tiny density fluctuations. A conservative picker may miss most true particles; an aggressive picker may collect a haystack of junk with a few needles inside. Either error can doom the reconstruction. The researcher is left with a familiar cryo-EM paradox: to know whether you have good particles, you would like a reconstruction; to obtain a reconstruction, you need good particles.

Phase plates and better detectors can ease this tension by making particles more visible at the outset, which improves both picking and alignment. So can careful specimen optimization that produces a cleaner background and a more uniform ice layer. But even with these advantages, the lower mass limit is not a fixed line. It moves with technology and with the ingenuity of sample preparation, and it moves unevenly across targets. A compact, rigid 80 kDa enzyme may be easier than a floppy 120 kDa multi-domain protein. A protein with a tight binder may be easier than a slightly larger protein alone. In cryo-EM, size is influential but not destiny.

Behind all the technical details lies a very human scientific question: what counts as an honest picture? When a small protein is solved with a scaffold or a binder, the map is not a simple window onto a native object; it is an image of a carefully constructed assembly designed to be visible. The result can still be immensely valuable, especially when it reveals the arrangement of active-site residues or the geometry of a binding pocket. But it asks for candor about what was added and what might have been constrained. The strongest work in this area tends to treat the added mass as a tool with a known footprint rather than as an invisible helper.

Small proteins also encourage a certain kind of experimental humility. They remind researchers that a beautiful reconstruction is not merely a reward for collecting enough data; it is a reward for a whole chain of decisions that kept the signal from being lost. When a project succeeds, it often does so through a series of small, almost unglamorous victories: a buffer tweak that reduces aggregation, a grid that produces more uniform ice, a binder that grips without distorting, an imaging condition that yields steadier contrast, an alignment strategy that resists being hypnotized by noise. The final map can look serenely authoritative, but behind it is usually an extended negotiation with a molecule that was never eager to be seen.

Hemoglobin, seen at 3.2 Å with a Volta phase plate, offered a clear proof that the small-protein barrier was not absolute. It did not make the barrier disappear. Instead it made it newly interesting: not a wall, but a frontier shaped by contrast, by computation, and by the behavior of proteins in a film of frozen water. For the proteins that vanish most readily, progress has often come from accepting a slightly paradoxical truth: to see something small, it can help to give it something larger to hold onto, provided one never forgets that the handle is not the object itself.

10.3 Inside the Frozen Cell

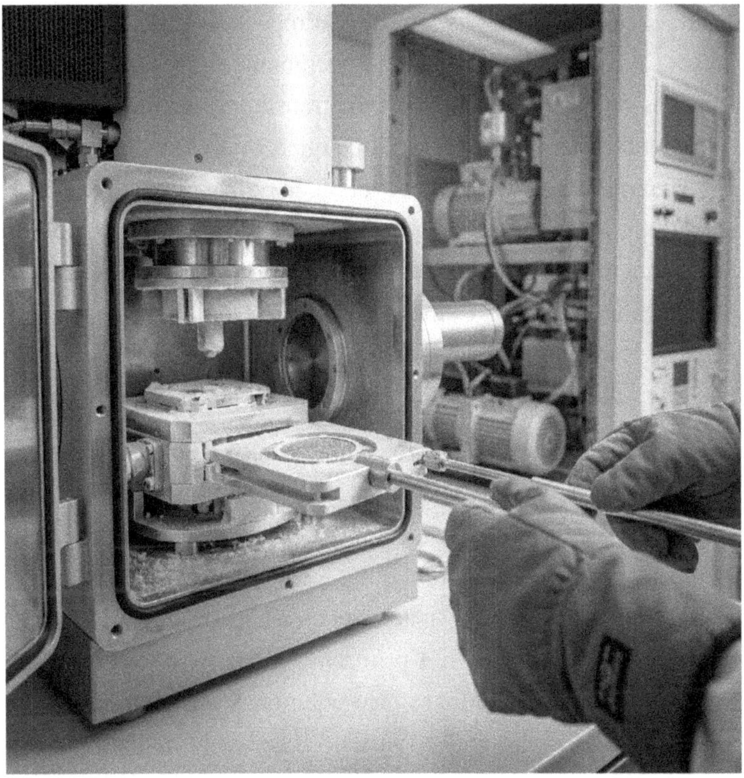

A purified protein, floating alone in a thin layer of frozen water, is a generous subject. It has been separated from its usual crowd, washed of many of the interactions that complicate its life, and presented as thousands of near-identical copies for the patient arithmetic of averaging. That isolation brings a kind of honesty: the structure that emerges belongs to a specific molecule in a specific biochemical state. Yet the same isolation can also be a distortion. Many molecular machines do not operate as solitary sculptures; they operate as neighbors. They lean against membranes, they queue at pores, they dock at scaffolds, they cluster in patches, they jostle for access to partners, and they often change their behavior depending on where they are placed.

Inside a cell, geometry is not a backdrop. It is part of the mechanism. A receptor embedded in a membrane senses different possibilities on each side. A ribosome translating near the endoplasmic reticulum can be physically coupled to a translocation channel, shaping what the nascent protein can do as it emerges. Even something as mundane as crowding—the fact that the cytoplasm is dense with macromolecules—can favor compact conformations and discourage others. Remove the protein from that environment and you may gain a clearer portrait while losing the very context that gives the portrait meaning.

Cryo-electron tomography, often shortened to cryo-ET, is built for this discomfort with isolation. It asks a different question from single-particle analysis. Instead of collecting images of thousands of separate particles and combining them into one high-resolution map, tomography keeps its attention on a particular piece of a particular cell. The aim is not primarily to perfect an average, but to reconstruct a three-dimensional volume: a snapshot of cellular architecture in a near-native, vitrified state, with membranes, filaments, and macromolecular assemblies preserved in place.

The central trick is both simple and technically demanding. A single 2D cryo-EM image is a projection: everything in the thickness of the sample collapses into one plane. Tomography regains depth by taking many projections from different angles. The specimen is tilted incrementally—a few degrees at a time—and an image is recorded at each tilt. The stack of images is called a *tilt series*. Computational reconstruction then combines these angled views into a 3D volume known as a *tomogram*. The logic resembles medical CT scans, but the scale, the fragility, and the limits are very different.

Those limits show up immediately in the way dose is budgeted. Electrons damage biological material. As discussed earlier, cryo-EM survives this by using low doses and by spreading the information across many particles that can be averaged. Tomography cannot spread the dose across many copies in the same way, because it is committed to one region of one cell. Yet tomography must record dozens of images at different angles, and each exposure consumes part of the sample's radiation tolerance. The dose has to be divided across the tilt series, meaning each image is noisier than a typical single-particle micrograph. The result is a familiar trade: a tomogram is rich in context but poor in signal. You can see where things are, but you do not automatically see what they are in atomic detail.

There is also a geometric tax. In an ideal world one would tilt from -90° to $+90^\circ$, collecting a complete set of views. In practice, the microscope stage and the specimen holder cannot reach those extremes without collisions and loss of stability. Many tilt series cover something like $\pm 60^\circ$ or $\pm 70^\circ$. This missing range produces a characteristic directional blurring in the reconstruction, often called the *missing wedge*: information that would have come from the unattainable angles is absent. The tomogram is therefore anisotropic, with some directions better resolved than others. Even without naming the mathematics, the intuition is clear: if you never see the specimen from certain directions, some details remain stubbornly uncertain.

If the method were only about tilting and reconstruction, it would already be challenging. The true gatekeeper, however, is thickness. Electrons do not pass through arbitrarily thick biological material without multiple scattering, loss of contrast, and severe blurring. Whole cells are often too thick for high-quality cryo-ET, especially eukaryotic cells with crowded interiors. The most biologically interesting regions—nuclear interiors, organelles, dense cytoplasm—are often the most inaccessible.

For a long time, this made tomography feel like a technique with a narrow habitat: thin bacteria, the edges of cells, naturally slender protrusions, or reconstituted systems such as membrane vesicles. It could show beautiful scenes, but it could not easily enter the cell's central rooms.

Cryo-focused ion beam milling, cryo-FIB, changed that practicality. A focused ion beam is a tool borrowed from materials science: a beam of ions, often gallium, that can sputter away material with nanometer-scale precision. In cryo-FIB, the specimen is kept frozen and the ion beam is used not to image but to *sculpt*. A thick frozen cell can be milled into a thin slab, a *lamella*, typically a few hundred nanometers thick, thin enough for electrons to traverse. This is not a gentle act. One is carving away most of the cell to expose an interior cross-section, all while trying to keep the remaining slice vitrified, uncontaminated, and mechanically stable.

The lamella is both an enabler and a new source of difficulties. Throughput is a constant pressure: milling takes time, and a single grid can contain many cells, each with its own idiosyncratic behavior under the beam. Contamination can accumulate as frost from residual water vapor, adding unwanted material to the surface. Milling can produce *curtaining*, a rippled texture caused by uneven sputtering, which degrades image quality and can masquerade as biological structure. Charging can build up on the specimen, distorting the beam. Lamellae can bend, crack, or deform, especially as they are transferred between instruments. Cryo-FIB has become an essential gateway to in-cell tomography, but it remains a craft where success depends on both engineering and patience.

Even when thickness is solved, there remains the problem of finding what you care about. Cells are large compared with the field of view used for high-resolution cryo-EM. Many targets are rare: a particular stage of a viral infection, a transient assembly at a membrane contact site, a few nuclear pores in a specific configuration. Searching blindly with the electron beam is inefficient and risks damaging the very areas one hopes to image.

This is where correlative light-and-electron microscopy, cryo-CLEM, enters as a practical guide. Fluorescence microscopy can mark specific proteins or structures with colored tags. Under cryogenic conditions, a fluorescent image can be acquired that indicates where, in a frozen grid, the target is likely to be. That map then guides FIB milling and tomography, turning a hunt into a targeted operation. Correlation is not perfect—fluorescent resolution is limited, and the relationship between the light microscope and the electron microscope involves alignment errors and distortions—but it can make the difference between repeatedly missing the target and consistently finding it.

Why go to such lengths? Because the reward is not merely a prettier picture of a familiar molecule. The reward is the chance to see how molecular machines are arranged as a community, and to treat proximity and placement as part of function. A striking example of this framing appeared in work on the nuclear periphery of HeLa cells in 2016, described as a kind of *molecular sociology*. The phrase is memorable because it captures something that textbooks often flatten. The nuclear envelope is not just a boundary; it is a busy border crossing. Nuclear pores, lamins, membrane proteins, and associated complexes form a landscape that organizes transport, mechanics, and signaling. Tomography can show not only the presence of these components but their spatial relationships: which complexes cluster, which align, which occupy specific neighborhoods. Even at modest resolution, such arrangements can change the kinds of questions one asks. Instead of treating a complex as a free-floating actor, one begins to ask what it is attached to, how it is oriented, and what local crowd it lives among.

Tomograms also reveal a kind of honesty about membranes and filaments that single-particle reconstructions often hide. A purified complex can be captured in exquisite detail, yet its placement in a membrane patch, or its alignment relative to a cytoskeletal filament, may be precisely what controls its activity. Tomography can show ribosomes studding a membrane, vesicles docked at specific sites, filaments weaving past organelles. Many of these scenes are recognizable even when the molecular detail is limited, in the same way that a city skyline is recognizable from far away. The promise is that some of those recognizable shapes can be pushed toward molecular detail by improved processing, better detectors, and clever averaging of repeated structures embedded in the tomogram.

Modern cryo-ET has been increasingly shaped by the realization that computation can recover more than the raw images seem to contain. Some of the same ideas that transformed single-particle analysis are now central to tomography. One is contrast transfer function (CTF) correction, which accounts for how the microscope modifies different spatial frequencies. In tomography this becomes more intricate, because each tilt image has its own defocus geometry and the specimen thickness means different parts of the volume can sit at different effective defoci. Another is refined alignment, not just of the overall tilt series but of the specimen itself as it subtly deforms or shifts under the beam during collection. Even the missing wedge, once a blunt limitation to be endured, can be partially compensated for in downstream analysis, especially when repeated particles are averaged in ways that combine their information.

The practical experience of looking at a tomogram can be initially disappointing to someone accustomed to the crispness of single-particle maps. A tomogram is grainy. The eye must learn a new kind of reading: membranes as continuous ribbons, ribosomes as dense beads, microtubules as hollow cylinders, large assemblies as irregular masses. Identification is often probabilistic, supported by shape, location, and biological expectation. This uncertainty is not a flaw so much as a faithful reflection of the constraints: low dose per image, limited angular range, and the complexity of the cellular environment.

Yet the very fact that tomography can keep the cell's architecture intact changes what becomes possible. Consider a membrane protein complex that is difficult to purify without altering its partners. In a tomogram, the complex can be seen embedded in its native membrane, with neighboring proteins and local curvature preserved. Or consider a large assembly that forms only in response to stress or infection. Tomography can capture it in the act, without the requirement that it survive purification. In these cases, the tomogram functions as a frozen field note, taken in situ rather than in a reconstituted laboratory setting.

The approach also has a moral dimension for those who care about biology as it actually happens. Purification is a kind of editing. It removes weak interactions, transient partners, and spatial constraints. Sometimes that editing is desirable, because it isolates a mechanism cleanly. Sometimes it turns a social molecule into a loner and then asks it to explain its behavior. Tomography cannot answer every question, and it often cannot yet deliver the atomic detail that makes single-particle cryo-EM so powerful. But it can restore something that is otherwise easy to lose: the sense that molecular machines are built not only to have a shape, but to occupy a place.

When cryo-FIB milling succeeds, when a lamella survives transfer and a tilt series accumulates without the sample collapsing under dose, the resulting volume carries an unusual intimacy. It is not an averaged archetype assembled from thousands of unrelated particles. It is a particular neighborhood, preserved at a moment, dense with structures that were never asked to leave home. In that frozen slice of cell, the molecules are not performing for the microscope. They are simply there, arranged in the ways that make living matter work.

10.4 Taking a Census in the Cellular Crowd

A tomogram can be strangely moving even when it is visually rough. Membranes arc and fold; filaments run like cables; dense particles cluster where you would expect the cell's busy places. Yet anyone who has stared at a tomogram for the first time recognizes the frustration that follows the awe. The scene is crowded, and the crowd is noisy. Many macromolecules are present only as faint blobs; others overlap; some sit at awkward angles where the missing wedge blurs them into uncertainty. The cell is intact, but the portrait is grainy.

This is the point where cryo-ET borrows back one of the most powerful ideas from single-particle cryo-EM: the idea that repeated observations, correctly aligned, can be combined into something sharper than any one observation alone. The trick is to do it without surrendering the very thing tomography was meant to preserve: context. The result is called *subtomogram averaging*. It sounds like a compromise, and it is, but it is also one of the most creative compromises in modern microscopy: a way to take a noisy three-dimensional cityscape and, by recognizing recurring buildings, learn their architecture in detail.

The basic unit is the subtomogram: a small cubic cutout extracted from a tomogram, centered on a suspected copy of some complex. If the complex appears many times in the volume—ribosomes are a classic example, but so are motors, coat proteins, and membrane-embedded assemblies—then one can extract many such cubes, rotate and shift them into a common alignment, and average them. Each cube is an imperfect view. Together, if the alignment is good, they can add up to a coherent map with much higher signal-to-noise ratio than the tomogram itself.

Subtomogram averaging is sometimes described as "single-particle analysis in 3D," but that phrase hides what makes it hard. In single-particle work the particles are already isolated in thin ice, and the images are 2D projections collected under carefully controlled conditions. In tomography, each particle is embedded in a thick, complex environment. It may be partly occluded by nearby structures. It may be deformed by local membrane curvature. The dose has been split across many tilts, so each view is noisy. On top of that, every particle is reconstructed with its own missing wedge, and the direction of that wedge depends on the particle's orientation in the cell. A simple average that ignores these anisotropies can smear the structure in unpredictable ways.

Yet the temptation to do it is irresistible, because the reward is precisely what many biologists want: not just a single purified structure, but a structure in place. A ribosome in a test tube is informative; a ribosome docked to a membrane translocon in a cell, surrounded by its neighbors, is a different kind of information. Subtomogram averaging is one of the few methods that can, in favorable cases, deliver both.

The first problem is simply finding the particles. Human eyes are good at spotting patterns, but they tire quickly, and they are not easily quantified. It is possible, and still common, to manually pick some structures in tomograms, especially when the object is large and distinctive. But manual picking does not scale, and it is vulnerable to a subtle bias: you tend to pick what you expect to see. That can become a self-fulfilling prophecy when the data are noisy. To take a census, one needs a method that is systematic enough to be trusted and flexible enough to work in diverse cellular neighborhoods.

A widely used strategy is *template matching*. The idea is to take a reference density—a rough model of the complex of interest—and slide it through the tomogram, computing at each position how well the local density matches the template under various rotations. Peaks in this correlation score become candidate locations of the complex. The approach is intuitive: if ribosomes have a characteristic shape, then regions that resemble that shape are likely ribosomes. In clean cases it works beautifully, turning a tomogram into a map of positions and orientations. It can also produce false positives, because noise and unrelated structures can sometimes resemble the template well enough to score highly, especially when the missing wedge has distorted the data in a consistent way.

Template matching therefore comes with a philosophical hazard. If the template is too detailed or too strongly expected, the search can become a hunt for your own assumptions. This is why practitioners often begin with low-resolution templates, or with templates that have been filtered and simplified, and why they insist on validation steps that ask whether the detected particles genuinely contain the expected structural features when averaged independently of the search template. The goal is not to avoid using prior knowledge—that is impossible in practice—but to prevent prior knowledge from silently becoming the result.

Once candidates are found, the next steps are extraction and alignment. Each candidate becomes a subtomogram cutout. The cutouts are then aligned to each other in 3D, refining rotations and translations so that corresponding features overlap. In principle, alignment is easier in 3D than in 2D because the subtomogram already has depth. In practice, the subtomogram's noise and anisotropy can make alignment fragile, and local cellular clutter can dominate the signal. Two ribosomes in different neighborhoods may have the same core structure but different surrounding densities, such as membrane nearby in one case and none in another. If the alignment algorithm is not carefully constrained, it can try to align the surroundings rather than the particle itself.

This is one reason subtomogram workflows often include masking: focusing the alignment on a region believed to belong to the particle and down-weighting the rest. Masking is powerful but also a choice. It forces the researcher to decide what counts as particle and what counts as context, a decision that can affect the result. In tomography, more than in purified single-particle work, that decision can be genuinely ambiguous. Is a membrane-embedded complex separable from the membrane, or is the membrane part of its functional shape? Is a bound factor part of the same "particle" for averaging, or a variable neighbor that should be excluded? The answers are biological as much as computational.

Averaging itself is straightforward in concept and treacherous in execution. When many aligned subtomograms are averaged, features that are consistent across copies reinforce each other. Random noise cancels out. The map sharpens. But features that vary—a flexible arm, a transient binding partner, a conformational change associated with function—will blur or disappear, just as heterogeneity blurs single-particle reconstructions. The tension returns: one wants both a sharper structure and an honest description of variability.

Classification can help. Subtomograms can be grouped into classes representing different states, and averages can be computed for each class. The challenge is again that the data are noisier than in single-particle work, and the missing wedge can bias classification toward differences in orientation rather than true structural differences. A dataset might accidentally separate into classes that correspond to whether particles were oriented with their missing wedge in one direction or another, rather than whether they were biologically distinct. Modern tomographic processing has invested considerable effort into modeling these sampling effects more accurately—including missing-wedge compensation, 3D-CTF correction, and refined alignment—so that classification becomes more about biology and less about geometry.

The aspiration that emerges from these workflows is sometimes called *visual proteomics*. The phrase captures a bold idea: treat the tomogram not only as a picture but as a dataset from which one can identify, count, and localize many molecular species in a cell. A census, in other words. The promise is enormous. Imagine being able to say not only that a particular complex exists in a cell, but how many copies there are, where they are, what orientations they adopt, and what fraction of them occupy each conformational state. At that point, cryo-ET becomes not merely descriptive but quantitative: a way of measuring molecular populations in their native environment.

This quantitative turn changes how one thinks about a tomogram. A tomogram is no longer simply an illustration to accompany a story about cell architecture. It becomes a coordinate system. Every detected complex has a position and an orientation. Those coordinates can be correlated with membranes, filaments, or other complexes. They can reveal clustering, exclusion zones, gradients, and preferred alignments. They can show whether a complex tends to sit at curved membranes or flat ones, whether it points in a particular direction relative to a filament, whether it forms ordered arrays or scattered distributions.

There is a very human satisfaction to this. The cell, which often feels overwhelmingly complex, begins to yield to a kind of counting that resembles ecology: not just what creatures exist in the forest, but where they nest, how they group, and how their distribution changes with conditions. The cell becomes legible as a populated environment.

A concrete example that many people have seen, even if they do not recognize it as such, is the ribosome-studded surface of the rough endoplasmic reticulum. In conventional cell biology images, that roughness is often shown as a stippled membrane. Tomography can capture the 3D arrangement of those ribosomes. Subtomogram averaging can then refine the ribosome structure in situ and, crucially, relate ribosome positions to the membrane and to associated machinery. Even when the final average is not at atomic resolution, it can reveal how ribosomes are oriented, how closely they pack, and whether particular factors appear consistently in the neighborhood. That begins to link a textbook cartoon—a membrane with dots—to a measurable architecture.

Viruses provide another compelling case because they build repeated structures in dense cellular contexts. Viral assembly often occurs at membranes, producing many copies of similar complexes arranged in patterns. Tomography can capture these patterns in a way that purified particles cannot, and subtomogram averaging can sharpen the repeating units. In fortunate situations, one can see both the individual subunits and their higher-order arrangement: the molecular and the architectural at once.

To do this responsibly, the field has had to become unusually serious about validation. A census is only useful if the counts are credible. Template matching can overcount by including false positives. It can undercount by missing particles in crowded regions or in orientations where the template correlates poorly because of anisotropic resolution. Classification can split particles into spurious groups driven by missing-wedge anisotropy. Even the definition of what constitutes a "particle" can vary. Does a partially assembled complex count? Does a complex bound to a factor count as the same species or a different one?

These questions do not have purely technical answers. They are, in part, choices about what biological categories are meaningful. But the technical community has developed statistical and methodological instincts to keep the choices from drifting into wishful thinking. One instinct is to test how robust detections are to changes in the template: does a different low-pass filter or a different initial reference produce the same peaks? Another is to split the dataset and ask whether independent halves produce the same averaged features, a discipline borrowed from single-particle validation. Yet another is to examine negative controls: regions where the complex should not exist, or templates for complexes that should not match, to estimate false-positive rates.

The phrase "template bias" has become a warning sign in this domain. It does not mean templates are forbidden; it means templates are powerful enough to create illusions if one is careless. A template is a hypothesis about what you are looking for. Template matching is then a hypothesis-driven search. The responsibility is to show that the subsequent averages contain information that was not simply imposed by the hypothesis. In a sense, subtomogram averaging demands a slightly different kind of honesty than single-particle work: not only "does the map fit the data?" but also "did the search strategy determine what was found?"

Software ecosystems have emerged that take these tomographic complications seriously. For example, emClarity has emphasized more accurate modeling of sampling, CTF effects, and anisotropy, with the goal of pushing cryo-ET closer to higher resolution and more reliable classification. The existence of such specialized tools reflects a mature recognition that tomography is not merely single-particle analysis performed on a thicker sample. It is its own statistical environment, with its own biases that must be modeled rather than ignored.

Even with excellent tools, subtomogram averaging retains a distinct flavor: it is an argument made from many weak witnesses. No single subtomogram is persuasive. A dozen are suggestive. Hundreds begin to speak clearly. Thousands can become remarkably sharp, provided the complex is rigid enough and repeated often enough. Yet the witnesses do not all tell the same story, and the differences among them can carry biological meaning. A "census" therefore becomes more than a count; it becomes a distribution. How many copies are present in each state? How does the distribution change when the cell is stressed, infected, or mutated? Does the complex form clusters or remain solitary? Does it adopt preferred orientations relative to cellular landmarks?

This is where the emotional appeal of in situ work becomes clear. It is not only about seeing the cell as a landscape; it is about seeing the cell as a population. The traditional high-resolution structure, magnificent as it is, can sometimes encourage a static view of life: one mechanism, one arrangement, one optimal configuration. A census encourages a probabilistic view. It invites questions that resemble physiology more than sculpture: what is typical, what is rare, what changes, what is spatially organized, what is conditional.

At its best, subtomogram averaging produces a peculiar kind of double vision. On one hand, there is the averaged structure: a sharpened portrait distilled from many copies. On the other hand, there is the return to the cityscape: the knowledge that each copy came from a particular spot, at a particular angle, in a particular neighborhood. The structure is not floating in abstraction; it has addresses. And once molecules have addresses, biology begins to look less like a list of parts and more like an organized crowd, where function can depend as much on who lives next door as on the shape of any single resident.

10.5 When the Microscope Starts Thinking Ahead

For much of cryo-EM's history, the limiting reagent was not liquid nitrogen or microscope time, but judgment. Someone had to decide whether a grid was worth loading, which squares to image, which holes had ice thin enough to be useful, whether the particles were real or merely dirt, whether the motion correction had worked, whether the first reconstruction was promising or deceptive. Expertise accumulated as a kind of taste: a fast, half-conscious ability to glance at a noisy micrograph and sense whether the experiment was on track.

Taste is powerful, but it does not scale. Modern cryo-EM produces data at a pace that can overwhelm a human operator, and access to high-end microscopes has broadened beyond a small guild of specialists. The field has therefore been drifting toward something that once would have sounded like science fiction: microscopes and software pipelines that evaluate their own output in real time, adjust their behavior, and increasingly share the burden of decision-making. The instrument begins to think ahead—not in the sense of having intentions, but in the more practical sense of anticipating what the experiment needs next.

The shift is easiest to appreciate by remembering how slow feedback used to be. A day of data collection could end with terabytes on a server but little clarity about whether anything usable had been captured. The true verdict might arrive days later after preprocessing, particle picking, and the first attempt at reconstruction. If the sample was damaged, the ice too thick, the orientations biased, or the microscope slightly misaligned, the discovery came late, and the precious microscope time was already spent. This delay shaped behavior: cautious operators collected conservative datasets, repeated what had worked before, and leaned heavily on familiar samples while new ones waited their turn.

Faster detectors and better automation have already shortened that lag, but the deeper change has come from software that treats the experiment as something that can be monitored and steered while it is happening. A tool such as Warp, described by Tegunov and Cramer (2019), embodies this attitude. Instead of waiting until the end of a session to process data, it can perform key preprocessing steps and quality evaluation as images are acquired. Motion correction and preliminary contrast transfer assessments can be carried out on the fly, and the operator can watch quality metrics evolve in real time. The practical consequence is surprisingly emotional: the experiment stops feeling like a blind leap and starts feeling like a conversation. If the ice is drifting toward thickness, if motion is worsening, if astigmatism creeps in, the signs appear early enough to act.

That capacity supports a kind of "faster failure," which is more valuable than it sounds. In a research culture that often rewards perseverance, failure tends to be recognized late. Real-time evaluation permits a different discipline: abandon a bad grid early, change conditions, or stop collecting once the data are clearly good enough. On an instrument where time is precious and demand is constant, the ability to stop intelligently can be as important as the ability to start.

Automation has also been transforming the earliest stages of a cryo-EM session: grid screening and selection. Traditionally, screening was a ritual of expertise. A skilled microscopist would load a grid, scan at low magnification, choose promising squares, then choose promising holes, then collect a handful of test images, mentally weighing ice thickness, particle concentration, aggregation, preferred orientation, contamination, and beam-induced motion. The process could be taught, but it was not easily standardized. Two experts might make different choices, and those choices could determine whether a project succeeded.

SmartScope (2022) represents a newer approach: deep-learning-driven automation for grid screening and documentation. Instead of relying purely on a person's memory and intuition, the system can identify and classify candidate imaging areas and record the decision path. This may sound bureaucratic, but it addresses a real problem. Cryo-EM experiments are expensive, and decisions made in the first hour can determine the outcome. Standardizing screening does not remove the need for expertise, but it does make expertise shareable. It also changes training. A novice can learn from consistent, annotated choices rather than from scattered verbal advice.

The most obvious arena for machine learning has been particle picking: the act of finding candidate particles in micrographs. Picking sounds mundane until one remembers that every downstream step depends on it. If the picked set is contaminated with junk, the reconstructions will be poor or misleading. If the picker misses a rare conformation, it may never appear in classification. If the picker is biased toward particles that resemble an existing template, it can quietly steer discovery away from the unexpected.

Here the contribution of Topaz, introduced by Bepler and colleagues (2019), has been widely influential. It uses positive-unlabeled learning, a strategy designed for situations where one has a modest set of positive examples but an enormous and ambiguous background. In cryo-EM, that is exactly the situation: one can label some true particles, but it is much harder to label every possible negative. By learning from a small number of curated positives and a large pool of unlabeled regions, Topaz reduces reliance on brittle heuristics and heavy manual labeling. The practical gains are speed and yield. The subtler gain is the possibility of reducing template bias, since the method can learn the appearance of particles directly from data rather than from a predefined, hand-crafted pattern.

These tools can be understood as individual helpers, but their real significance appears when they are connected into a loop. Picture an experiment where screening is automated and documented (SmartScope), where preprocessing and quality evaluation occur as each movie arrives (Warp), and where particle picking begins while the microscope is still collecting (Topaz). The operator does not have to wait for the end of the session to learn whether particles are present, whether the motion correction is behaving, or whether the dataset is dominated by ice contamination. Decisions that once depended on gut feeling can now be guided by near-immediate evidence.

This changes what "experience" means. Experience becomes less about private intuition and more about choosing thresholds, defining acceptable ranges, and knowing which warnings are meaningful. It can also make cryo-EM more accessible. A laboratory without a resident cryo-EM virtuoso can still run competent sessions if the software scaffolding is strong, and if expert oversight can be applied at the right moments rather than at every moment. This is one reason cryo-EM has been spreading beyond flagship centers into a wider range of institutions.

Yet accessibility has a shadow side. A human expert has biases, but those biases can be discussed and challenged. A learned system has biases too, and they can be harder to see. If a particle picker is trained mostly on clean, well-behaved particles, it may systematically miss unusual appearances: partially assembled complexes, flexible states, rare binding partners, or damaged yet biologically informative intermediates. If a screening system learns that "good" holes tend to have certain textures, it might reject conditions that are genuinely novel but visually unfamiliar. Automation can turn taste into code, and code can freeze taste into a rule that quietly shapes what the community studies.

This raises a new kind of responsibility: the responsibility to ask what the system has been taught to value. In cryo-EM, value is often expressed as a narrow set of metrics: thinner ice, higher contrast, less motion, more particles per micrograph, sharper Thon rings, better early reconstructions. These are sensible goals, but they are not always aligned with discovery. Some of the most interesting samples are unstable, heterogeneous, and rare. They may never produce "perfect" micrographs. A pipeline that always seeks the highest immediate quality may steer attention away from these awkward cases, reinforcing a subtle conservatism.

The deeper debate is about reliability and reproducibility when machine learning systems prioritize what to image or what to keep. Traditional cryo-EM workflows were hard to reproduce in part because decisions were implicit: a person chose one hole and not another, kept one class and discarded another. Automated systems can make workflows *more* reproducible by recording decisions and applying them consistently. But they can also make errors more systematic. A flawed model can be applied at scale, producing a large and consistent mistake. When results are wrong in a human-driven workflow, they can be idiosyncratic; when results are wrong in an automated workflow, they can be industrial.

That is why human oversight remains essential even as tools become more capable. Oversight does not mean refusing automation; it means keeping the power to question it. In practice, this can involve training pickers on diverse examples rather than on one pristine dataset, periodically auditing what is being rejected, and testing whether conclusions depend on a particular model or threshold. It can also involve cultural habits: encouraging people to report failures and oddities rather than hiding them, and rewarding careful validation rather than only speed.

One of the most intriguing prospects is adaptive data collection: the idea that the microscope could change its behavior based on what the data are showing. Even without naming a specific "self-driving" system, the ingredients are increasingly visible. If preprocessing can estimate motion and image quality in real time, the system can preferentially collect from areas that are performing well. If early reconstructions suggest preferred orientation, the system could adjust tilt strategy to diversify views. If particle concentration is low, the system could move to different grid squares or adjust targeting. If enough high-quality data have been collected to reach the desired resolution, the system could stop, saving time and dose for other projects.

The phrase "microscope starts thinking" can sound like an invitation to overstate what is happening. No current instrument is discovering biology on its own. What is changing is more practical and, in its own way, more profound: the distribution of attention. Cryo-EM used to require sustained expert attention across hours of collection and days of processing. Increasingly, attention is concentrated at key decision points, supported by automated monitoring elsewhere. This is how many complex technologies mature. Airplanes did not become safer by eliminating pilots, but by creating systems that monitor, warn, stabilize, and assist, while still leaving ultimate responsibility with a person.

There is also a scientific payoff beyond convenience. Faster, more standardized pipelines can make it easier to compare experiments across labs, because data quality and preprocessing become less dependent on local quirks. They can also accelerate iteration. If a sample behaves poorly, one can learn that quickly, adjust conditions, and test again, rather than waiting a week to discover that the first attempt was doomed. In a field where sample preparation is often the true bottleneck, speed of iteration can translate directly into scientific progress.

A historical anecdote from the early days of cryo-EM, passed down in many laboratories, is the memory of "waiting for the film to come back." Before direct detectors and automated pipelines, images were recorded on photographic film. Developing and scanning the film took time, and the delay meant that mistakes were often discovered late. The romance of film has its charm, but the inefficiency shaped behavior: it rewarded caution and made risky experiments expensive. The move to digital cameras made immediate viewing possible, but the real transformation has been the move from immediate viewing to immediate *assessment*. The microscope is not only a camera now; it is part of a measurement system that can judge its own performance while it measures.

This creates a new partnership between human and machine. The human contribution is still uniquely valuable where biology is messy: deciding what question is worth asking, recognizing when a strange dataset might be a discovery rather than a failure, noticing when a "bad" grid reveals an unexpected assembly, understanding which compromises are acceptable for the biological claim. The machine contribution is relentless consistency: never tiring, never losing focus, applying the same standards to every micrograph, and surfacing patterns that a person might miss in the flood.

The most interesting future is therefore not one where the microscope replaces expertise, but one where expertise is amplified and redistributed. A skilled cryo-EM practitioner may spend less time clicking through holes and more time thinking about experimental design, validation, and interpretation. A newcomer may be able to achieve competent data collection sooner, guided by tools that encode hard-won lessons. Large facilities may be able to support more users without sacrificing quality, because the early stages of screening and preprocessing become less dependent on a handful of experts.

There will also be moments when the partnership must be renegotiated. A system that optimizes for standard quality may need to be deliberately overridden to pursue an unusual target. A picker that performs brilliantly on one class of particles may fail on another. An adaptive collector may decide that a dataset is "good enough" when the biological question demands subtler distinctions. These are not problems to be solved once and for all; they are the ordinary tensions of working with tools that have preferences.

Cryo-EM began as a way of freezing a fleeting molecular world into something the eye could inspect. It is gradually becoming a way of running an experiment that inspects itself as it unfolds, catching problems early, guiding effort where it is most productive, and freeing human attention for the parts of science that cannot be automated: curiosity, skcpticism, and the willingness to take a promising anomaly seriously.

www.ingramcontent.com/pod-product-compliance
Lightning Source LLC
LaVergne TN
LVHW010712061126
851857LV00014B/1155
* 9 7 9 8 2 4 0 8 3 6 3 7 4 *